# Incremental Sheet Forming Technologies

# Incremental Sheet Forming Technologies

## Principles, Merits, Limitations and Applications

Ajay and Ravi Kant Mittal

CRC Press
Taylor & Francis Group
Boca Raton London New York

CRC Press is an imprint of the
Taylor & Francis Group, an **informa** business

First edition published 2020
by CRC Press
6000 Broken Sound Parkway NW, Suite 300, Boca Raton, FL 33487-2742

and by CRC Press
2 Park Square, Milton Park, Abingdon, Oxon, OX14 4RN



*Library of Congress Cataloging-in-Publication Data*

Names: Ajay, author. | Mittal, Ravi Kant, author.
Title: Incremental sheet forming technologies : principles, merits, limitations, and applications / Ajay, Ravi Kant Mittal.
Description: First edition. | Boca Raton, FL : CRC Press, 2021. | Includes bibliographical references and index.
Identifiers: LCCN 2020020214 (print) | LCCN 2020020215 (ebook) | ISBN 9780367276744 (hardback) | ISBN 9780429298905 (ebook)
Subjects: LCSH: Sheet-metal work.
Classification: LCC TS250 .A38 2021 (print) | LCC TS250 (ebook) | DDC 671.8/23--dc23
LC record available at https://lccn.loc.gov/2020020214
LC ebook record available at https://lccn.loc.gov/2020020215

ISBN: 978-0-367-27674-4 (hbk)
ISBN: 978-0-429-29890-5 (ebk)

Typeset in Times
by Deanta Global Publishing Services, Chennai, India

# Contents

# List of Figures

# List of Tables

# Preface

Sheet material forming is one of the most widely used manufacturing methods for producing parts from sheet material, required for the production of high-quality components. A large variety of sheet forming processes are in vogue in industry. Many of these sheet deformation processes, used in making parts for automobiles, aircraft and other products, use very complex equipment and dies. The research in the sheet material deforming processes and the introduction of new techniques is enduring and complex, for objectives like better productivity, at lower cost with high precision, for complex and unusual applications, like in medical and space applications. In-depth exploration of this area shows meager published literature and the absence of a book for the new technology. This encouraged the authors to undertake the responsibility of disseminating knowledge in the form of the present book on incremental sheet forming.

This book provides an overview of the extensive research done in sheet material forming, whose outcome was perfecting a new technique – incremental sheet forming (ISF). The area is under-explored and the outcomes are worth the research effort. Experimentation, modeling and simulation, the powerful tools for developing new concepts and getting valuable information on the process, led to the novel process development for sheet material forming. Since the material available on ISF is scattered across patents and research publications and not available in one place in book form, the authors recognize their ethical responsibility to compile, share and spread the knowledge accumulated and technology developed, with students, researchers and industry people, to draw the benefits of the new technology and gain technical competence in the upcoming area.

ISF is a novel sheet forming process that eliminates the use of expensive dies and reduces tooling costs for the manufacturing of complex sheet material parts, which makes it ideal for manufacturing prototypes and low-volume production as compared to other traditional sheet material forming processes. ISF also finds suitability for producing sheet material components of old machinery, which are otherwise very difficult and non-economical to produce due to the unavailability of forming dies. The ISF process is a die-less process that uses a simple forming tool which is made to move along a controlled tool-path for forming the sheet locally layer by layer, into the desired shape. This technique is a major breakthrough in the sheet material forming processes without using a punch and die.

ISF is a novel and flexible forming technique for producing highly customized sheet material components economically. Hence, it becomes important to understand the concept and usefulness of the ISF technique. To facilitate detailed learning of ISF, this book introduces the principles, benefits and limitations of all variants of the ISF technique along with their applications. It further discusses forming machinery and other hardware required to accomplish this novel sheet material forming technique. The implementation of computer-aided design and manufacture for ISF is then introduced. The influence of variation of ISF process parameters is elaborated with suitable examples. Details of variations in forming forces, formability, surface finish, thickness reduction, thinning limit and forming fracture limit diagrams of ISF are discussed. In the end, applications, challenges and future scope of various variants of ISF are delineated. This book is an attempt to provide a comprehensive and substantial compendium to enhance the readers' knowledge of the ISF technique.

Although it is hard to present an extensive and comprehensive account of all approaches of sheet material techniques due to the rapid technological enhancements in this field, the authors have, however, made their best attempt at making this book comprehensive and informative by describing variants and advancements of this novel technique. The authors have reviewed substantial literature and utilized practical knowledge gained by several years of work and research experience to write this book. The book concentrates on simple engineering rather than complex models and mathematical simulation, to provide the student, researcher, academician, practicing engineer and shop floor technician with usable information in simple lucid language.

This book is intended for both academia and industry. Postgraduate students, Ph.D. students and researchers in universities and institutions, who are involved in the advancements in material working, will find this compilation useful. Based on this book, it would be possible to offer a one-semester course for postgraduate students in mechanical engineering or manufacturing engineering, with supplementary material on modeling, analysis, design and simulation ISF. Similarly, the book will help manufacturers of regular sheet material parts as well as those manufacturing sheet material parts for the biomedical, automotive and aerospace applications, or who are involved in in-house R&D for improving their products and designing new products.

The authors acknowledge the professional support received from CRC Press and express their gratitude for this opportunity.

Readers' observations, suggestions and queries are welcome.

**Ajay**
**Ravi Kant Mittal**

# Acknowledgments

The authors express their hearty adulation and gratitude to Ms. Cindy Carelli (editor), CRC Press, for giving her consent to publish our work and to transmit an arranged scientific format of this manuscript. Her suggestions and directions were helpful in intensifying the work and certainly improved the content of the manuscript. She undoubtedly imparted great and adept experience, and her systematic and methodical staff immensely helped the authors to complete the manuscript. The authors are grateful to CRC Press for showing their interest in publishing this book on a cutting-edge area of current manufacturing techniques and the book will have great future prospects.

The author, Dr. Ajay, wishes to thank Prof. Vishal Gulati and Prof. Ravi Kant Mittal for their unreserved guidance, valuable suggestion, thought-provoking discussion, motivation and encouragement in nurturing this work. Prof. Gulati is a wonderful and intelligent person who is ever ready to help his students in every possible manner. The present work is a testimony to his alacrity, inspiration and ardent personal interest, taken by him during the current project in its present form. Prof. Mittal is the epitome of simplicity, forthrightness and strength and is a role model for the author. It has been a blessing for the author to have many precious moments under the guidance of the perfectionist at the acme of professionalism. The author also wishes to express his gratitude to his parents, Sh. Jagdish and Smt. Kamla, for their true and endless support. They have made him able to walk tall before the world while sacrificing their happiness and living in a small village. Words are not enough to appreciate the world's best parents. He cannot close these prefatory remarks without expressing his deep sense of gratitude and reverence to his dear wife Sarita for her understanding, care, support and encouragement to keep his morale high throughout the period of work. The author feels extremely happy to express his sincere appreciation to his brother Parveen for his understanding, care, support and encouragement. No amount of words can ever quantify the love and gratitude the author feels in thanking his daughter, little SejalRathee, who is the world's best child.

The author, Prof. Mittal, feels fortunate to come in contact with Dr. Ajay, when he joined K. R. Mangalam University. He found Dr. Ajay to be an intelligent hard-working intellectual with zeal to achieve. It has been a pleasure to work closely with him on this book project. The author would also like to acknowledge all his seniors, peers, associates and students whom he came across in his 45-year career and who aided him. They all have been wonderful people who always moved along with him. Prof. Mittal acknowledges the backing he has received from his life partner

Mrs. Abha Mittal and all their children and their life partners for their untiring, timely and curative care extended by them 24/7, all these years.

Finally, the authors dedicate this work to the divine creator and express their indebtedness to the "ALMIGHTY" for gifting them the power to yield their ideas and concepts into substantial manifestation. The authors believe that this book will enlighten readers about each feature and characteristic of incremental sheet forming techniques.

**Dr. Ajay**
**Dr. Ravi Kant Mittal**

# Authors' Biographies

**Dr. Ajay** is currently serving as an associate professor in the Mechanical Engineering Department, Faculty of Engineering and Technology, SGT University, Gurugram, Haryana, India. He received his Ph.D. in the field of Incremental Sheet Forming from Guru Jambheswer University of Science & Technology, Hisar, India. He completed his B.Tech. (Hons.) and M.Tech. (Distinction) at Maharshi Dayanand University, Rohtak, India. His areas of research include incremental sheet forming, additive manufacturing, advanced manufacturing and optimization techniques. He has over 25 publications in international journals of repute and refereed international conferences. He is supervising three Ph.D. and two M.Tech. scholars. He has a total of eight years of experience in teaching and research. He has won several proficiency awards over the course of his career, including merit awards, best teacher awards and so on.

At SGT University, he is an advisor of the Association of Engineers and Technocrats (AET) and has also authored many in-house course notes, lab manuals, monographs and invited chapters in books. He has organized a series of faculty development programs, workshops and seminars for researchers, UG and PG level students. He teaches the following courses at the graduate and postgraduate level: Manufacturing Technology, Advanced Manufacturing Processes, Material Science, CAM, Operations Research, Optimization Techniques, Engineering Mechanics, Computer Graphics, Design of Experiments and Research Methodology and so on. He is associated with many research, academic and professional societies in various capacities.

**Prof. (Dr.) Ravi Kant Mittal** began his academic career at Birla Institute of Technology and Science, Pilani (BITS Pilani), India, one of the top universities of India, in 1975 and retired from there as Senior Professor in 2018. He headed the offshore campus of BITS Pilani at Dubai, UAE, as Director and was also the Vice-Chancellor of a private university, K.R. Mangalam University, Gurugram, India, for a term. In his academic endeavor, he has introduced, developed and taught a wide spectrum of courses in mechanical engineering, computer science and new courses in emerging and interdisciplinary areas, at the graduate and postgraduate levels, including product design, mechanical design, reliability engineering, control systems, systems modeling, robotics and intelligent systems, mechatronics, MEMS and nanotechnology. He is a life member of the IEEE and ACM.

Prof. Mittal is author/co-author of over 80 technical papers and has co-authored 2 popular textbooks: *Elements of Manufacturing Processes*, PHI Learning; and *Robotics & Control*, McGraw-Hill. He was editor of two international conference proceedings and has also authored many in-house course notes, lab manuals, monographs and invited chapters in books. Seven students have completed their

Ph.D.s under his supervision, and his research interests include robust robot design, robot-path planning, micro-electromechanical systems (MEMS), nanotechnology, software engineering, software testing and e-waste. He has guided more than 60 M.E./M.S. dissertations and numerous undergraduate projects/theses.

At BITS Pilani, he held a series of leadership positions, including Director (Special Projects), Deputy Director Administration, Dean, Academic Registration and Counselling Division and founder Chief, Computer Assisted Housekeeping Unit and IEEE Student Branch Counsellor, amongst others. He was instrumental in establishing the BITS Alumni Association (BITSAA) and was its Founder President. He received a B.E. (Hons.) and M.E. in Mechanical Engineering, and a Ph.D. degree from BITS Pilani, India. He obtained the highest rank in order of merit in his M.E. and was awarded the Institute's Gold Medal.

# List of Abbreviations

| | |
|---|---|
| **3D** | Three-dimensional |
| **2D** | Two-dimensional |
| **AA** | Aluminum alloy |
| **ADSIF** | Accumulative double-sided incremental forming |
| **AISI** | American Iron and Steel Institute |
| **C-ISF** | Conventional ISF |
| **CAD** | Computer-aided designing |
| **CAM** | Computer-aided manufacturing |
| **CDD** | Charged coupled device |
| **CMM** | Coordinate measuring machine |
| **CNC** | Computer numerical control |
| **CRCA** | Cold rolled close annealed steel |
| **DC** | Direct current |
| **DDQ** | Deep drawing quality |
| **DOF** | Degree of freedom |
| **DSIF** | Double-sided incremental forming |
| **EDD** | Extra deep drawing |
| **FE** | Finite element |
| **FFL** | Fracture forming limit |
| **FLC** | Forming limit curve |
| **FLD** | Forming limit diagram |
| **FPG** | Fasting plasma glucose |
| **GCA** | Grid circle analysis |
| **HA-ISF** | Heat-assisted ISF |
| **HSS** | High speed steel |
| **ISF** | Incremental sheet forming |
| **NC** | Numerical control |
| **ORB** | Oblique roller-ball |
| **PAO** | Pulsed anodic oxidation |
| **PCL** | Polycaprolactone |
| **PID** | Proportional integral derivative |
| **PLA** | Polylactic acid |
| **PTFE** | Polytetrafluoroethylene |
| **PVC** | Polyvinyl chloride |
| **R&D** | Research & design |
| **RPM** | Rotation per minute |
| **RTD** | Reaming-tapping-drilling |
| **SPIF** | Single-point incremental forming |
| **STL** | Standard triangle language |
| **TPIF** | Two-point incremental forming |
| **VMM** | Video measuring machine |
| **VRB** | Vertical roller-ball |

| | |
|---|---|
| **VWACF** | Varying wall angle conical frustum |
| **VWAPF** | Varying wall angle pyramidal frustum |

**Nomenclature**

| | |
|---|---|
| $\alpha$ | Wall angle |
| $\alpha_f$ | Final wall angle |
| $\alpha_p$ | Pre-form wall angle |
| $\varepsilon_1$ | Major strain |
| $\varepsilon_2$ | Minor strain |
| **μm** | Micron/micro-meter |
| **cSt** | Centistokes |
| $F_x$ | Forming force along *x*-axis |
| $F_y$ | Forming force along *y*-axis |
| $F_z$ | Forming force along *z*-axis |
| $L_m$ | Measuring length |
| **mm** | Millimeter |
| $R_a$ | Average roughness |
| $R_z$ | Maximum roughness |
| $t_f$ | Final thickness |
| $t_i$ | Initial thickness |
| $t_p$ | Pre-form wall thickness |
| **Yr** | Roughness spread from reference line |

# 1 Introduction and Need for Advanced Sheet Forming Techniques

## 1.1 INTRODUCTION

Since the advent of the human race many processes, tools and machines have evolved to manufacture products to make life easier and more enjoyable. In the twentieth century, many evolutions have taken place in an effort to meet the manufacturing challenges created by newer requirements, materials, tools and cost constraints, and the quest continues. The developments and advancements in the aero-space, automobile, marine, medical and many other sectors often require make-to-order manufacturing, and newer technology is required to meet this challenge. Such manufacturing systems need autonomous, flexibly integrated and adaptable processes that can respond quickly to dynamic requirements with real-time shop floor control of the process. Incremental sheet forming is an evolving modern manufacturing process for these situations, one of the buzz areas of research in sheet material manufacturing, for the economic manufacturing of custom objects. Manufacturing is a complex activity of converting raw materials into finished products with desired properties, shape and size, using a wide variety of processes, machinery, tools and other equipment with or without automation, the use of computers, robots, sensors and other special equipment; it requires people with diverse skills, knowledge and expertise in different disciplines. The effectiveness and efficiency of a manufacturing method depend on the flexibility and agility of the process as well as its feasibility for a particular application.

To carry out a manufacturing activity we require five Ms, viz. *M*en, *M*aterials, *M*achines, *M*oney and *M*ethods. Methods refer to *manufacturing methods* used for making the desired products economically and quickly, with anticipated quality and adaptability to ever-changing needs and technological developments. In addition, knowledge of the various materials, tools, machine tools, accessories, manufacturing processes, metrology and testing instruments for checking materials or products is essential for understanding manufacturing processes.

## 1.2 MANUFACTURING PROCESSES

The manufacturing processes or techniques for producing desired products or components are conventionally grouped into five basic production processes based on the handling of the material required to convert it into a component by giving it the desired shape, size, accuracy, finish and properties. These are:

- Casting processes
- Material removal processes
- Forming/deformation processes
- Joining processes
- Special processes

The basics of these processes are briefly described here to refresh the concepts of the different methods of manufacturing, the machines and tools required for each of them and the terminologies involved.

### 1.2.1 Casting Processes

Casting is one of the oldest known methods for shaping materials. In casting processes, the material is melted and liquid material is poured into a cavity (called a mold) of the shape and size of the required product where it is allowed to solidify. After solidification, the product, called "castings", is extracted from the mold. Casting processes are widely used to produce products of small sizes to very large sizes, with very intricate shapes and with strength and rigidity frequently not obtainable by other manufacturing processes.

The casting process has low production costs but is a labor-intensive process, requiring other operations to get the finished product, with limitations of dimensional accuracy and surface finish. Special casting processes, like the die-casting process, investment casting process, vacuum-sealed molding process and shell molding process, overcome these limitations.

### 1.2.2 Material Removal Processes

*The class of manufacturing processes* that use cutting or machining to remove excess material for producing the required shape, size, accuracy and finish are known as material removal processes or more commonly as machining processes. The removal of unwanted material from the raw material is done by using specific shaped *cutting tools*, operated manually or by machines called *machine tools*. With machining processes, it is possible to manufacture components with very low dimensional tolerances (accuracy and precision) that cannot be obtained by other methods of manufacturing but generate a lot of waste material from the removed portions as *scrap*. Typical examples of products made by machining processes include gears, automobile parts, wood furniture, nuts and bolts, etc. Examples of common machining processes include *drilling*, *turning*, *milling*, *grinding*, *boring*, *knurling*, *facing*, *reaming*, *chamfering*, etc. Machining is used as a secondary or subsequent process for many of the other manufacturing processes like casting, forming, etc.

### 1.2.3 Forming Processes

Manufacturing processes for giving the desired shape and size to raw material by permanently deforming it using force, pressure or stresses like compression, tension, shear or their combinations are known as forming processes. Unlike casting or machining, in forming, the material is neither melted nor removed. Forming processes exploit a property of materials – the ability to flow plastically (or deform) in the solid state (without melting) – when subjected to suitable stresses. The material is deformed to the desired shape and size with almost no wastage. The required force is generally high in forming processes, and the tools and equipment needed are quite expensive. Commonly, a machine for forming processes is known as a *press*, and forming tools used for the same are called a *die and punch*. Typical examples of the products manufactured by forming processes include rails, kitchen utensils, wires, drink cans, collapsible tubes and so on. No material wastage, a high cost of tooling and fast and easy large-quantity production of components are some of the characteristics of forming processes. Brittle material cannot be deformed plastically and, hence, cannot be formed. However, semi-brittle material can be formed using hot working (forming above the recrystallization temperature).

Blanking, punching, bending, stretching, rolling, forging, drawing, extrusion, spinning, etc., are some common forming processes. Forming processes are classified into two groups – bulk forming processes and sheet forming processes. The classification of forming processes is given in Figure 1.1.

Bulk forming involves severe deformation of material resulting in massive shape change. The surface-area-to-volume ratio of the workpiece is relatively small. Bulk forming is mostly done in hot working conditions. Some common bulk forming operations are forging, rolling, drawing and extrusion. Sheet forming processes are discussed in Section 1.3.

### 1.2.4 Joining Processes

The category of manufacturing processes where two or more workpieces are joined together to produce the required shape and/or size of the product is known

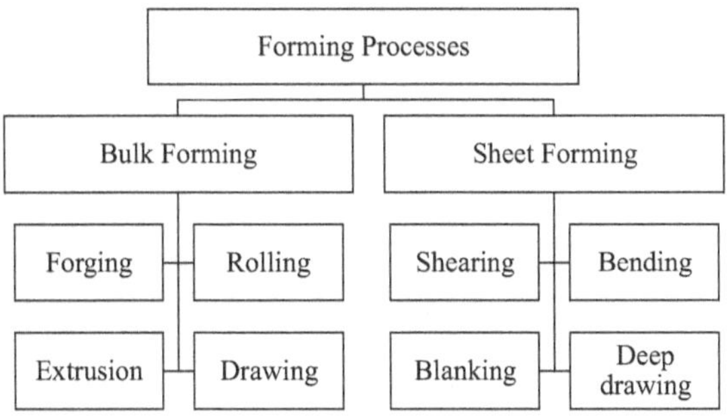


**FIGURE 1.1** Classification of forming processes.

as joining processes. Joining processes (welding, mechanical fasteners, adhesive, etc.) permit complex shapes to be constructed from simpler parts and have a wide domain of applications. Before joining, the components may be manufactured by any of the processes discussed above. The joint, of two or more parts, can be *permanent, semi-permanent* or *temporary*. Permanent joining can be done by fusing the material of parts together. For this kind of joining, material is locally heated or melted (with or without applying pressure) and some filler *material* may be used. Examples of permanent joining processes include *welding, brazing, soldering*, etc. Temporary joining of components can be done by using rivets, nuts-bolts, screws, etc. Adhesives are also used to make temporary joints. Joining processes are widely used in *fabrication.*

### 1.2.5 Special Manufacturing Processes

Many other special technologies of manufacturing are in vogue for specific materials, products, situations and applications such as 3D printing, powder metallurgy and electrochemical machining, etc. These are also called *non-conventional manufacturing processes.*

### 1.2.6 Selection of Manufacturing Process

It is possible to manufacture a component or product by more than one manufacturing process. How to select the most suitable manufacturing process for manufacturing a component? The parameters that influence the selection of a suitable process include the technical viability of a process for the product, the volume of production, the expected quality, the properties required, economy, energy and material efficiency. Besides these, infrastructure, convenience, planning, scheduling, inventory control, etc., are also important criteria for selecting an economical process for production.

## 1.3 BASICS OF SHEET MATERIAL WORKING

Sheet material forming started in the early days of humanity's evolution and is a sub-class of material forming processes. Products made from sheet material are all around us and are widely used. An extensive range of products is manufactured with the use of sheet materials such as parts of cars, trains, aircrafts, agricultural equipment, office equipment, furniture, house appliances, computers, machine components and beverage cans to name a few. Material economy and the resultant reduction in weight (being lightweight) and lower production cost, high productivity, use of semiskilled labor, versatility of shapes and sizes and a high degree of precision have rendered sheet metal processes indispensable for the economical mass production of products. Sheet material manufacturing produces parts that typically have high strength, good surfaces and accurate tolerances. A brief review of the basics of sheet material working processes is done first, before introducing the need for advanced sheet forming processes like the incremental sheet forming (ISF) process.

### 1.3.1 Sheet Materials

A material shape that has a high surface-area-to-volume ratio is called a sheet or plate. The dividing line between a sheet and plate is based on thickness. As a rule of thumb, a sheet is 3 mm or less in thickness, while a plate is over 3 mm in thickness. Sheet material work stock, used for sheet material processes, is the most commonly used material form, usually produced by rolling and available as coils, sheets and strips.

### 1.3.2 Sheet Material Selection

It is important to understand the mechanical behavior of sheet material before discussing sheet material working processes. For a required component, sheet material selection is a crucial part of product design. It affects the design, material cost, manufacturing process, manufacturing cost and performance of the product. The following factors affect the selection of the sheet thickness and material for a given product:

- Required application of product – shape, size, strength and weight constraints
- Cost of material and manufacturing
- Required properties of product like strength, corrosion resistance, etc.
- Assembly with other parts
- Availability

Material selection primarily depends on products' strength and properties requirements. An important and desirable property for sheet material forming is a large plastic deformation before necking. Grain size, structure and orientation are also important for sheet metal working. Grains will affect the properties of the metal as well as the surface finish. Table 1.1 gives a summary of important material characteristics and their significance for sheet forming operations.

Aluminum alloys and low-carbon steels are the most commonly used sheet metals because of their low cost, generally good strength and formability. Other commonly used sheet materials are Cold rolled close annealed steel (CRCA), galvanized iron, stainless steel, copper, tin, lead, zinc and plastic, etc. A wide range of alloys of ferrous and non-ferrous metals are used for specific applications.

### 1.3.3 Sheet Thickness Selection

Ideally, the sheet material process should not produce a change in the sheet thickness of the sheet material product produced but, in some situations, the product may require different thicknesses at different sections of the component. Structural simulation of the part may be performed to determine the required sheet thickness for a given material for the desired product properties.

### 1.3.4 Sheet Temperature during Forming

The temperature of the sheet material determines its ductility and hardness, and these in turn influence the forming force. Sheet material is usually formed at room

**TABLE 1.1**
**Material Characteristics and Their Significance for Sheet-Forming Operations**

| Characteristic | Significance |
|---|---|
| Elongation | The capability of the sheet material to stretch without necking and failure is known as elongation. Larger elongation of material results in higher formability. |
| Yield-point elongation | Yield-point elongation is typically observed with mild-steel sheets; results in depressions on the sheet surface; can be eliminated by temper rolling.<br>Some metals such as low-carbon steels and aluminum–magnesium alloys may experience yield-point elongation. This yielding of the material may produce stretcher strains or Lueder's bands [1] and spoil surface finish. |
| Formability | Formability is the ability or ease of a sheet material to undergo permanent shape change without failure by necking or tearing. The greater the plastic deformation before fracture, the greater the sheet's formability. Formability depends on properties of material such as yield strength, strain hardening rate, strength coefficient, grain size, ductility and so on. As the temperature of sheet material is increased, the yield strength and strain hardening rate are reduced whereas ductility is increased. Several methods for predicting the formability of sheet material are discussed in Chapter 6. |
| Strain hardening | Strain hardening refers to the fact that when a material is deformed in some area, dislocations occur in the microstructure. As these dislocations pile up, they tend to strengthen the metal against further deformation in that area. Thus, the strain is spread throughout the sheet. However, at some points in the material the strain suddenly localizes and necking develops, leading to fracture of the sheet in the necked region [2].<br>The strain-hardening coefficient, therefore, reflects how well the metal distributes the strain throughout the sheet, avoiding or delaying localized necking. The higher the strain-hardening coefficient, the more the material will harden as it is stretched and the greater will be the resistance to localized necking. Necks developed in the material harm the surface appearance and affect the structural integrity. Therefore, a high strain-hardening exponent and strain-rate sensitivity exponent are desirable. |
| Anisotropy (planar) | Anisotropy is an important factor in sheet forming. Anisotropy is the directional variation of mechanical properties – the material reacts differently to stress applied in one direction than if the same stress applied in a different direction. A sheet is isotropic if its properties are the same in all directions. The cupping test can be used to determine anisotropy.<br>Anisotropy is commonly present in cold-rolled sheets because of preferred rolling orientation. It causes defects like earing in deep-drawn products and the same can be reduced or eliminated by annealing which also lowers the strength of the sheet material. |
| Anisotropy (normal) | Anisotropy determines the thinning behavior of sheet during stretching. It is important in deep drawing processes. The anisotropy coefficient is derived from the ratio of the plastic width strain to the thickness strain.<br>A material with a high plastic anisotropy has a greater "thinning resistance". In general, the higher the anisotropy coefficient is, the better is the material deformation in drawing operations. |

*(Continued)*

**TABLE 1.1 (CONTINUED)**
**Material Characteristics and Their Significance for Sheet-Forming Operations**

| Characteristic | Significance |
|---|---|
| Grain size | Grain size determines surface roughness of a stretched sheet. The coarser the grain is, the rougher is the appearance (like an orange peel). It also affects material strength and ductility. |
| Residual stresses | Residual stresses are typically caused by non-uniform deformation during forming and may result in distortion of the part. This can also lead to stress-corrosion cracking. Residual stresses can be reduced or eliminated by stress relieving. |
| Springback | Springback occurs due to elastic recovery of the plastically deformed sheet after removing the forming force. It may cause distortion of the part and loss of dimensional accuracy. It can be controlled by techniques such as over bending and bottoming of the punch. |
| Wrinkling | Wrinkling is caused by compressive stresses in the plane of the sheet. It can be useful in imparting stiffness to parts by increasing their section modulus and can be controlled by proper punch and die design. |

temperature (called cold forming); however warm or hot working of sheet materials is possible and offers specific advantages.

*Cold forming* is generally done at room temperature, i.e. below the material's recrystallization temperature. Compared to hot forming, some of the advantages are closer dimensional tolerances, good surface finish, higher strength and hardness of the component because of strain hardening, no additional equipment requirements and lower costs. Cold forming requires a higher forming force and more powerful forming machines and puts limits on the deformation without failure. Cold forming processes, in general, are better suited to the large-scale production of parts.

*Warm forming* is done at temperatures just above room temperature but below the recrystallization temperature of the material. Compared to cold forming, due to the higher temperature of the material, more plastic deformation is possible with less forming force but additional heating equipment is required. Compared to hot forming, warm forming requires less heat energy and gives better precision and dimensional control of components, better surface finish and greater life of tooling.

*Hot forming* involves heating the material to be formed above its recrystallization temperature but below the melting point, i.e. the sheet temperature is 0.5 to 0.75 times the melting point. Since the temperature of the material is above the recrystallization temperature, no strain hardening takes place. The higher temperature of the material increases its ductility, reduces hardness and more plastic deformation can be obtained without failure, with lower forming force. Higher working temperatures reduce the life of forming tools (punch and die), give lower dimensional accuracy and produce poor surface finish, and extra heat left in the material after working promotes grain growth, leading to poor mechanical properties of the material. In hot forming, the handling of the hot work piece and maintaining its temperature are difficult and cumbersome.

### 1.3.5 Punches and Dies

Forming is conventionally carried out by using a punch and die, which are made to produce specific components. The mechanism of operation of a punch and die to form the sheet is shown in Figure 1.2. The die has the negative shape of the contour and the punch has the positive contour of the shape to be given to the sheet. A raw flat sheet is held over the die (or clamped), as shown in Figure 1.2a, and then the punch is forced into a sheet, during the working stroke, giving the desired shape to the sheet as shown in Figure 1.2b. On the return stroke, the punch is moved out and the formed sheet is removed from the punch by a stripper plate, giving the finished product, Figure 1.2c.

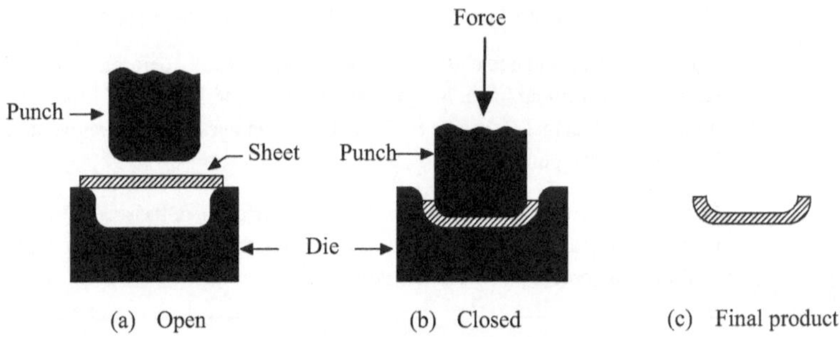


**FIGURE 1.2** Operation of punch and die for sheet forming: (a) open, (b) closed, (c) final product.

*Clearance*: In sheet forming operations, the blank size, the die size and the punch size are designed according to the final size of the desired component and the gap between the punch and die is known as the clearance as shown in Figure 1.2. The clearance is an important parameter for the correct forming of the sheet and depends on many factors such as material properties, sheet thickness, desired component thickness and so on.

## 1.4 SHEET MATERIAL FORMING OPERATIONS

Conventional sheet material forming operations, to give shape to the sheet, are performed by a large variety of methods. Sheet material forming operations can generally be classified into shearing and forming operations as shown in Figure 1.3.

### 1.4.1 Shearing Operations

The conventional sheet shearing operations include cutting, punching, blanking, piercing, notching, trimming, nibbling and many more. The schematics and characteristics of some of these operations are shown in Table 1.2 along with their descriptions.

### 1.4.2 Forming Operations

Forming is the most widely used sheet operation for manufacturing various sheet material components. Based upon the design of the component and the material

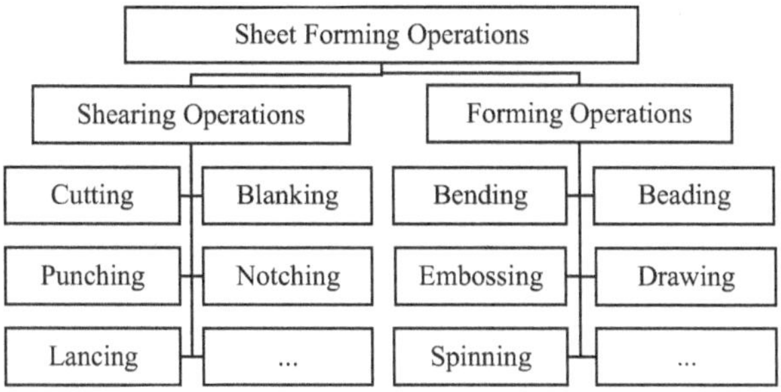


**FIGURE 1.3** Classification of sheet forming operations.

used, a variety of sheet forming processes have evolved, each with application and characteristic. The common sheet forming operations include bending, beading, hemming, drawing, spinning, etc. Some of these sheet forming operations are briefly described below.

*Bending*: Bending is one very common sheet metal forming operation wherein the sheet material is strained to give a bend to get the desired shape. As the sheet material is bent, its outer surface experiences the stretching, comes under tension, while the inner surface is compressed. Schematics of two common sheet bending setups are shown in Figure 1.4.

*Beading and hemming*: Beading, hemming, seaming, curling and similar other operations are forming operations done on the edge of the sheet for specific requirements like giving strength to the sheet edge, making the sheet edge smooth or joining two sheets at the edge. All these are variants of bending operations applied to the sheet edge. Figure 1.5 shows some of these sheet edge operations.

*Embossing*: Embossing is an operation in which sheet metal is drawn to shallow depths using male and female matching dies. This operation is carried out usually for the purpose of stiffening sheet panels or for making items like number plates, name plates, jewelry, etc. [3]. Figure 1.6 is a schematic of the embossing operation.

*Drawing*: The process of shaping a sheet blank into a three-dimensional hollow shape component without any appreciable change in sheet thickness is known as a *drawing process*. The process involves forcing the blank into the die cavity with a punch, which exerts sufficient force on the sheet and the sheet is drawn into the die opening. The sheet material deforms and takes the shape of the cavity between the die and punch [4]. The process of drawing a sheet to form a cup shape is shown in Figure 1.7.

The material being drawn must possess a combination of ductility and strength so that it does not rupture under the stresses it is subjected to, when the sheet is pushed into the die [5].

A considerable amount of compressive stress develops in the flange portion of the blank being drawn, and this may cause wrinkling of the sheet if the sheet thickness and clearance are not correct. To prevent wrinkling of the sheet a blank holder may be used to hold the blank firmly until the material is completely drawn.

In a single-stage drawing operation, it is possible to obtain a diameter 1.8 to 2 times less than that of the initial blank diameter. A large force is required for deep drawing, and if the applied force increases beyond the material rapture strength, the sheet will be fractured [6].

**TABLE 1.2**
**Characteristics of Common Sheet Shearing Operations**

| Operation | Schematic | Characteristics and Applications |
|---|---|---|
| Shearing/cutting | 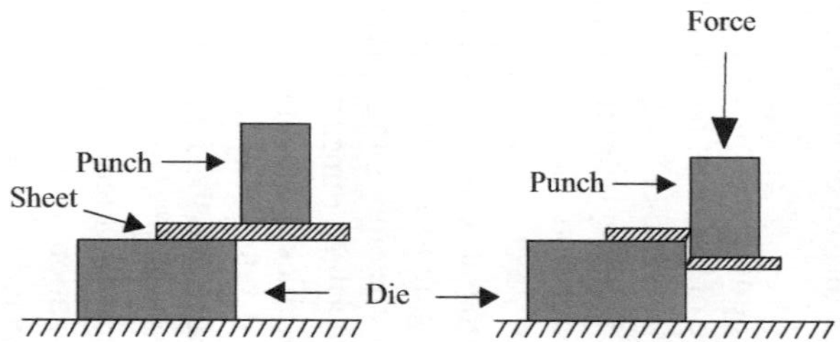  Shearing/cutting operation | Most basic sheet operation is cutting of the sheet by shearing action to get blanks of required sizes, for producing desired components. |
| Punching | 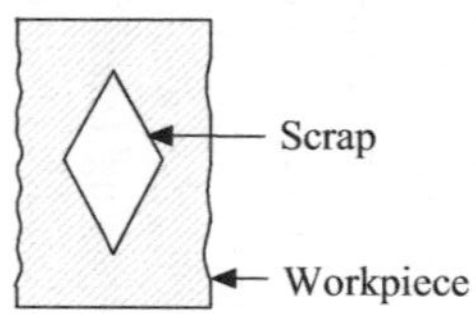  Punching operation | Punching or blanking is a process in which the punch removes a portion of the sheet from the larger piece or a strip of sheet material. If the removed piece is discarded, the operation is called punching, whereas if the removed piece is the useful part and the rest is scrap, the operation is called blanking. |
| Blanking | 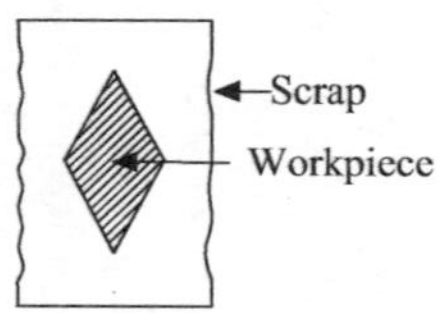  Blanking operation | |

(*Continued*)

**TABLE 1.2 (CONTINUED)**
**Characteristics of Common Sheet Shearing Operations**

| Operation | Schematic | Characteristics and Applications |
|---|---|---|
| Notching | 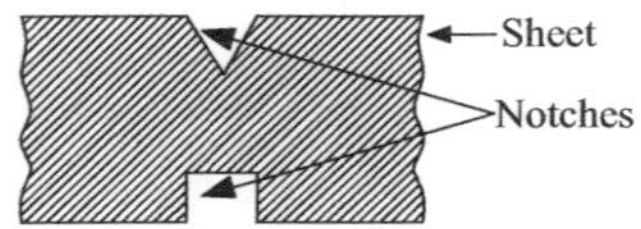<br>Notching operation | It is an operation in which a specified small amount of metal is cut from the edge of the blank. A notch can have any shape. |
| Lancing | 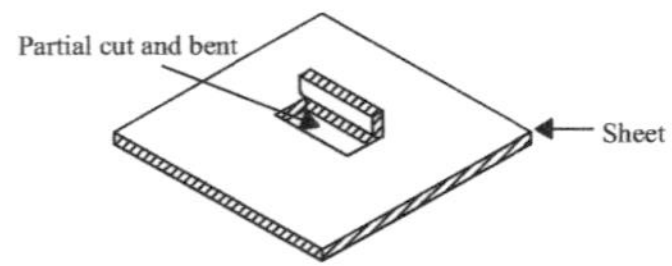<br>Lancing operation | The cutting of a sheet through part of its length and bending the cut portion up or down. |

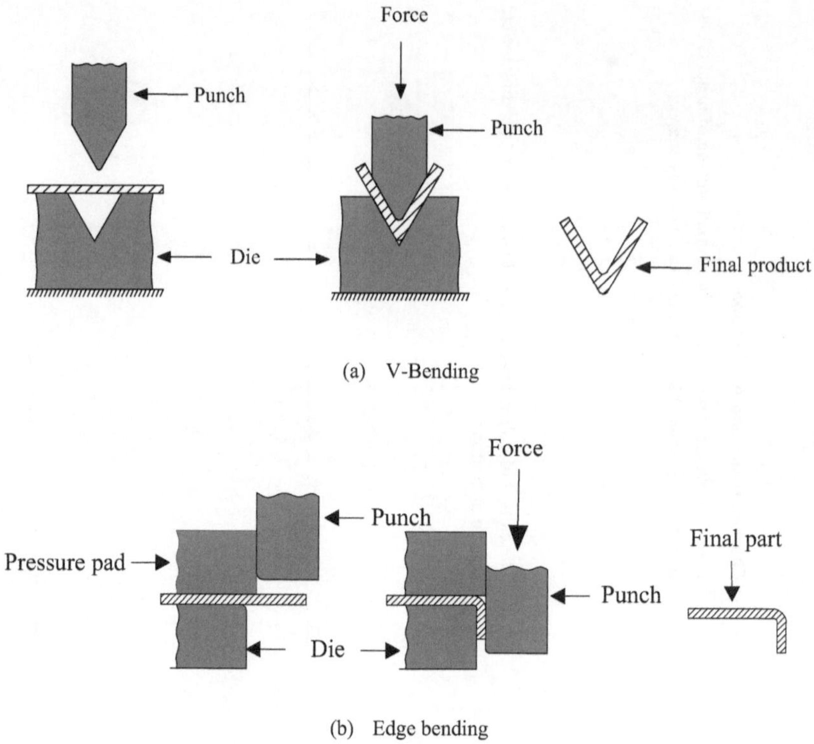


**FIGURE 1.4** Bending operations: (a) v-bending, (b) edge bending.

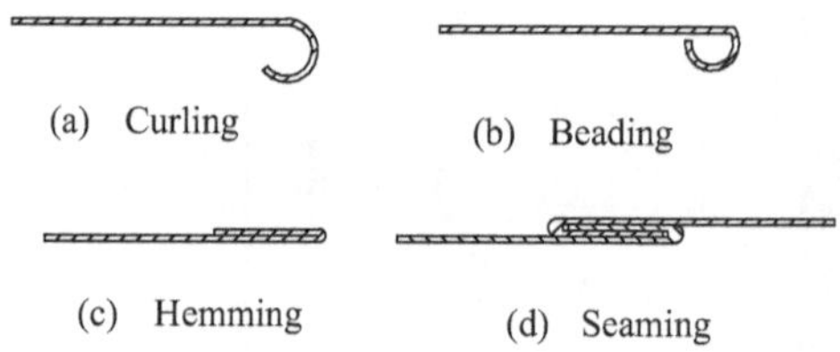


**FIGURE 1.5** Some sheet edge operations: (a) curling, (b) beading, (c) hemming, (d) seaming.

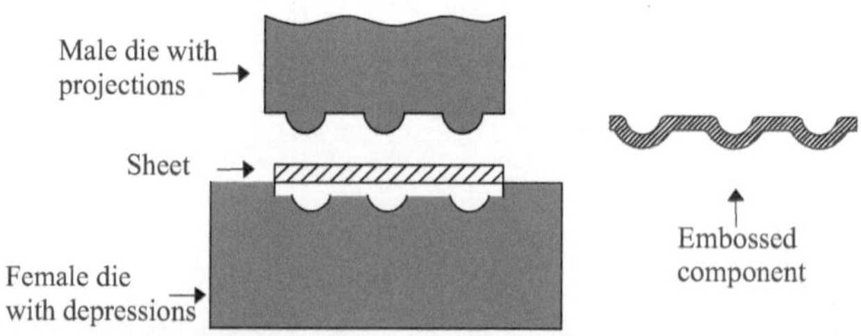


**FIGURE 1.6** Embossing operation.

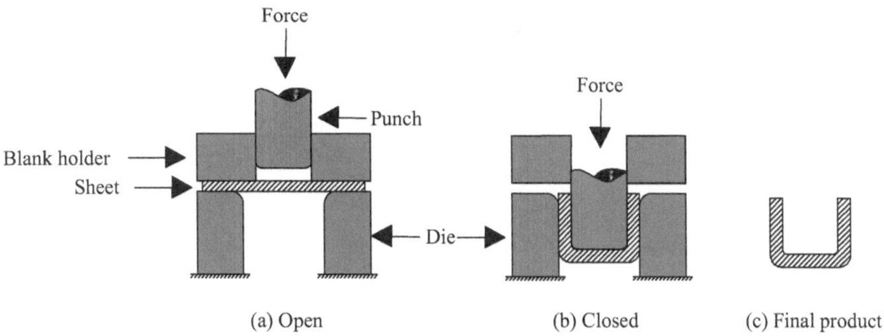


**FIGURE 1.7** Drawing operation: (a) open, (b) closed, (c) final product.

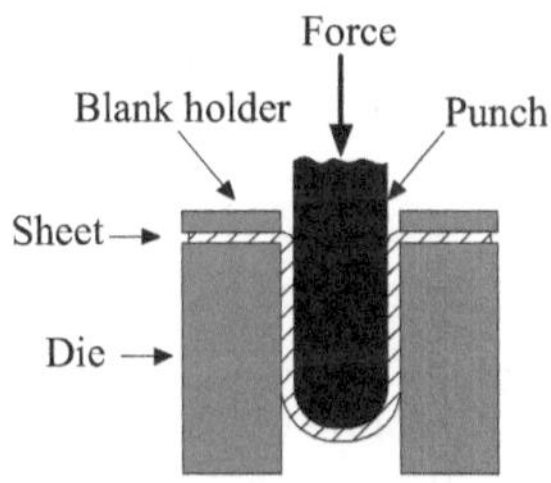


**FIGURE 1.8** Deep-drawing operation.

Shallow parts with relatively simple shapes, high production rates and high tooling and equipment costs are characteristics of the drawing process. Some lubricant, applied over the face of the blank, is generally used in sheet drawing operations to reduce friction and, hence, drawing force.

*Deep drawing*: The drawing process is considered "deep drawing" when the depth of the drawn part exceeds its diameter. Deep drawing is a widely used sheet forming process and is carried out in stages. The first stage of deep drawing is called cupping, similar to drawing process, with a pressure pad holding the blank on the die while the punch moves in and takes the blank into the die cavity (see Figure 1.8). The metal is deformed plastically and drawn over the edge till the hollow shape is formed. The pressure pad helps in minimizing the wrinkles on the surface of the part. Larger depths can be produced with repeated deep drawing. If the clearance between the die and the punch is less than the sheet thickness, then ironing occurs.

*Spinning*: Sheet material spinning is the process of producing axially symmetric components by rotating the clamped sheet at high speed and pressing it against a mandrel (or former) of the desired shape. The mandrel is fixed on the spindle of the lathe machine and the sheet is held in position using the tailstock or a similar arrangement. As the sheet and mandrel rotate, the operator applies localized force on the sheet using a special forming tool, causing the sheet material to flow progressively towards the mandrel, as illustrated in Figure 1.9, and ultimately the sheet assumes the shape of the mandrel. The deformations involved are a combination of bending

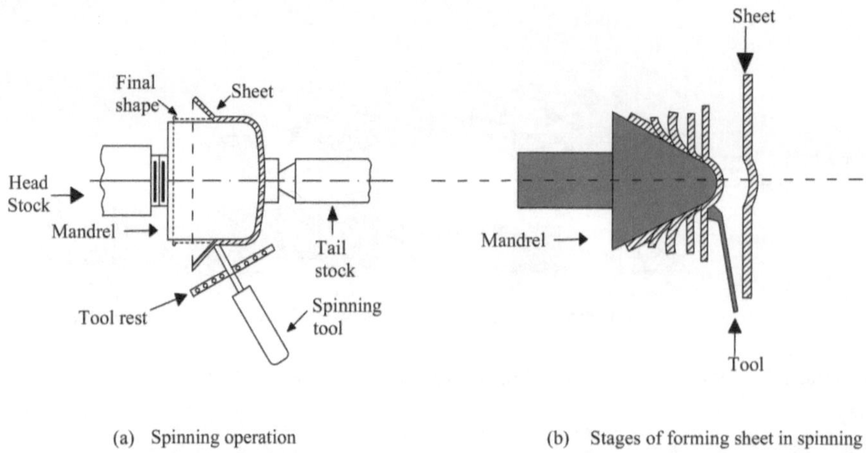


**FIGURE 1.9** Spinning process for sheet forming: (a) spinning operation, (b) stages of forming sheet in spinning.

and stretching mechanisms. The success of the process depends on the skill of the operator in guiding the sheet material to deform it to the shape of the mandrel [3].

Axially symmetric shapes, low tooling costs, high labor costs and a good surface finish are characteristics of the spinning process. This process is most suitable for shaping small or large hollow parts from ductile metals and alloys. Typical examples of the products made by this process include cooking utensils, bowls, lamp reflectors and the bells of some musical instruments.

### 1.4.3 Other Sheet Forming Operations

There are many other special sheet forming processes that have evolved for specific applications. The characteristics and applications of some of these are summarized in Table 1.3.

## 1.5 NEED FOR SPECIALIZED SHEET FORMING

The conventional sheet forming processes, discussed above, require custom-designed expensive tooling systems (punch and die) for each component to be manufactured by sheet forming and are, therefore, economically feasible only for mass production of the component. If only one or a few components are to be manufactured, conventional methods are highly uneconomical.

Hence, there is always a need for the development of innovative flexible manufacturing processes for small-quantity production and rapid prototyping of metal and non-metal components. The new global competitive environment is driving attention to such processes. In the case of parts made from sheets, a new class of forming processes denoted as "incremental sheet forming" processes is gaining importance due to its capabilities of meeting the variable requirements of manufacturing sheet material components.

**TABLE 1.3**
**Special Sheet-Material Forming Processes and Their Characteristics**

| Forming Process | Process Characteristics and Applications |
|---|---|
| Explosive forming | The pressure generated from an explosive charge underwater is used for forming the sheet placed over a die.<br>It is suitable for extremely large components like boat hulls and aircraft components with simple shapes. It has relatively low tooling costs but high labor costs, low-quantity production and long cycle times. It is used for batch type production. |
| Magnetic pulse forming | The high-intensity electromagnetic pulse is used for forming electrically conductive materials like aluminum and copper.<br>It is used for shallow forming, bulging of tubes and embossing operations on relatively low-strength materials. It avoids wrinkles, requires special tooling and has high cycle time. |
| Peen forming | Small round steel shots are blasted over the sheet to deform the sheet to give it the shape of a component. It is a die-less process.<br>It is suitable for shallow contours on large components and requires high equipment costs. |
| Roll forming | Sheet is passed through a set of shaped rollers to progressively change the shape of the sheet by very small amounts. It is suitable for constant cross section and long sheet parts with simple or complex cross sections. It generates good surface finish of the component and is used for mass production. It assumes lower production costs, high tooling costs and high production rates. |
| Stretch forming | Sheet blank is clamped at the boundary, and stretched and forced against the die to give it the shape of the die.<br>It is suitable for large components with shallow contours like an aircraft fuselage, wing panel, etc., at low production rates. Tooling and equipment costs increase with the increase in parts' size. The thickness of the sheet is reduced because of the stretching mechanism involved. |
| Super plastic forming | The sheet is formed at elevated temperatures to get the maximum elongation at plastic level. It is suitable for creating large and complex components in a single operation. Good surface finish and close dimensional precision, long forming times and low production rates are characteristics of it. |
| Coining | It is a severe metal-squeezing operation in which the flow of metal occurs only at the top layers of the material and not throughout. Material is squeezed in closed dies with pressures as high as five to six times the strength of material, for the purpose of producing coined parts.<br>Different designs with very fine details can be created on each side of the workpiece using this process. |

The ISF process is characterized by progressively applying plastic deformation to sheet material, with a forming tool that is maneuvered by a computer numerical control (CNC) action on a milling machine or by an industrial robot. The forming tool performs the incremental localized deformation of the sheet while moving along a predefined trajectory, descending a small step in each contour. The schematic of ISF is shown in Figure 1.10.

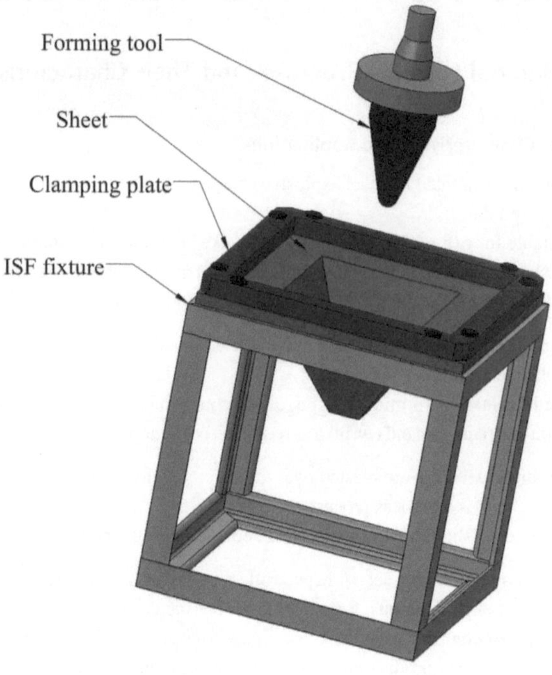


**FIGURE 1.10** Schematic representation of ISF.

The well-known and established spinning process described above in Section 1.4.2 is the origin of the ISF process. In spinning, the sheet is progressively deformed in order to produce axisymmetric parts. The sheet is rotated at high speeds and a form tool (or a roller) is moved over the sheet to progressively deform it to take the shape of the mandrel. In ISF, the sheet blank is mounted in a sheet holder and the forming tool is pressed into the sheet in a controlled contour to deform the latter and produce the desired shape. Unlike in spinning, the sheet is not rotated, a mandrel is not required and non-axis symmetric components can be produced by ISF.

## 1.6 ORGANIZATION OF THE BOOK

This book provides a detailed overview of the newly developed and innovative tool-based ISF processes. It is based on the extensive experimental research done on the ISF processes, and their input, output and process parameters. In each chapter, an extensive state-of-the-art review and claims made by the research community are discussed, analyzed and summarized. The presentation includes a comprehensive analysis of the analytical, numerical, and experimental modeling of the ISF process as well as examples of benchmark and industrial sheet parts made with metals, non-metals and polymers. The principles, historical timeline, current status of research, advancements, applications, key issues and scope of future work have been presented for the readers. However, readers with a keen interest in the specific topic or development should follow the references given at the end of each chapter, for more details.

The material in the book is organized into eight chapters, as follows:

*Chapter 1, Introduction and Need for Advanced Sheet Forming Techniques,* begins with manufacturing needs and presents the basics of different sheet forming processes and explains the need for advanced and flexible forming processes to meet the requirement of rapid prototyping and batch size production. The basics of ISF are also presented. This chapter also gives the outline of the book.

*Chapter 2, Incremental Sheet Forming: A Die-Less Approach*, describes the historical timeline of developments in ISF methods. The working principle, classification, advantages, limitations and key issues of the ISF process are also introduced briefly.

*Chapter 3, Incremental Sheet Forming Process*, describes the various types of ISF processes along with detailed descriptions, schematics, advantages and limitations of each forming method. Forming strategies, forming tools, forming machinery and forming temperatures are also discussed. The current state of research and the working principle of each process variant are also explained.

*Chapter 4, Machine Tools and Forming-Tools for Incremental Sheet Forming Process*, explains the machines and other hardware needed to accomplish this die-less process. Various parameters and criteria that influence the selection of forming machinery, forming tools, use of lubricants and computer-aided designing/computer-aided manufacturing (CAD/CAM) software are also delineated.

*Chapter 5, Incremental Sheet Forming Parameters*, provides the technological capabilities and particular drawbacks of the ISF process in terms of input parameters. The impact of these parameters, viz., workpiece parameters, machine parameters and process parameters, is discussed. The current state of research on input parameters of ISF is also presented to define the appropriate input parameters on an industrial scale. Furthermore, various tool path strategies have been elaborated with their potential uses that can help researchers and readers to understand the state-of-the-art techniques of ISF processes.

*Chapter 6, Responses of Incremental Sheet Forming Process*, addresses various output parameters (or responses) of ISF to determine the criterion for setting the values of input parameters for different applications. The methods of measuring the output parameters are also explained in this chapter, and the present state of research on various responses of ISF processes is also included.

*Chapter 7, Applications and Challenges of Incremental Sheet Forming Processes*, focuses on a variety of applications of the ISF process in medical, automotive, aerospace, architecture, prototyping, manufacturing, etc. The challenges and issues that limit the applicability of this process have also been discussed in this chapter. The proper understanding and knowledge of challenges can help researchers and industrial personnel to enhance the applicability and sustainability of the process on an industrial scale.

*Chapter 8, Conclusions and Future Scope*, concludes the book based on analysis exhaustively presented in the previous chapters of the book. The future possibilities of this die-less forming technology are extremely promising and have been discussed at the end of this chapter.

## REFERENCES

1. Danilov, V.I., V.V. Gorbatenko, L.B. Zuev, D.V. Orlova, and L.V. Danilova. "Luders Deformation of Low-Carbon Steel." *Steel in Translation* 47, no. 10 (2017): 662–668.
2. Ghosh, A.K. "Tensile Instability and Necking in Materials with Strain Hardening and Strain-Rate Hardening." *Acta Metallurgica* 25, no. 12 (1977): 1413–1424.
3. Parashar, B.S.N., and R.K. Mittal. *Elements of Manufacturing Processes*. PHI Learning Pvt. Ltd., 2002.
4. Groover, Mikell P. *Fundamentals of Modern Manufacturing: Materials Processes, and Systems*. John Wiley & Sons, 2007.
5. Lenard, John G. *Metal Forming Science and Practice: A State-of-the-Art Volume in Honour of Professor JA Schey's 80th Birthday*. Elsevier, 2002.
6. Schuler GmbH. *Metal Forming Handbook*. Springer Science & Business Media, 1998.

# 2 Incremental Sheet Forming

## *A Die-Less Approach*

## 2.1 INTRODUCTION

In this modern era, the international market is accepting an alarming growth of obsolescence in the use of technology to manufacture components due to cutting-edge and customer-oriented requirements. Economical and customized products have become a basic requirement in the manufacturing sector for companies to upgrade their business [1]. Batch-type production requires technologies that can produce components whose cycle time and set-up cost are significantly lower. Incremental sheet forming (ISF) has the capacity to produce sheet material components at lower cost due to the flexibility and elimination of costly dies [2]. ISF evolved in the beginning of the current century and is still developing as a forming methodology that has proved its significance in the manufacturing and design of customized sheet material components. Many recent studies [3–6] have demonstrated that ISF techniques have the capacity to trigger the next revolution in the field of advanced sheet forming methods. ISF has attracted global attention in the field of sheet forming processes. In addition, ISF is a viable and net-shape manufacturing process which allows a negligible amount of waste material on the blank edges. Undesirable edges of blank can be eliminated with a simple cutting operation. ISF has attracted the attention of researchers due to its flexibility and economical tooling. ISF is also suitable for forming non-symmetrical geometries without using expensive dies for manufacturing complex components of sheet metal [7]. As an example, replacing the fuselage parts of old aircraft is a major issue in the aerospace sector due to the unavailability of forming dies and punches of these components. These kinds of problems can be overcome by a process like ISF which is a purely die-less forming process. It requires minimum energy to form the components. It also significantly reduces the tooling costs as compared to other conventional sheet material processes where expensive dies are used to produce any product [8]. Hence, multi-variety components in small batches can be manufactured at low cost with ISF technology which prevents the limitations of traditional sheet material forming processes.

Furthermore, sheet material is squeezed locally by the forming-tool which results in a reduction in the forming-force and an increase in the formability of the process. The forming-force has a great impact on the precision and fracturing of the sheet material. Moreover, the stress level and forming-forces are directly related to each other. Stress is related to the plastic strain, which allows the calculation of the structural integrity of the formed part. Hence, it becomes customary to study these die-less techniques which find application in the fabricating of prototypes for the aerospace and biomedical sectors [9]. ISF is a die-less process that uses a simple spherical tool which moves along the CNC-controlled tool-path. It is based on deforming the sheet locally layer by layer depending upon the geometry of the target component.

## 2.2 HISTORY AND DEVELOPMENT

This section presents an image of the history and gradual development of the ISF technology. The last several decades have seen a rising attentiveness in a special category of forming technologies recognized as ISF. Different processes are characterized by the name "*incremental forming*" which means that only a minor area of the component is under local deformation at any instant of time during forming. A large number of processes like traditional blacksmith's hammering and hammer-forging techniques are covered under this definition. Although rolling may also be regarded as an incremental process, this has not been accepted by the researchers. Spinning has been regarded as a specific incremental process in order to form the sheet into batch-type products like household cooking items.

Spinning and ISF are similar techniques with some fundamental differences. To explore the history of ISF techniques, it becomes customary to acknowledge the thin line between ISF and spinning techniques. Spinning refers to clamping the sheet (workpiece) on a revolving mandrel and feeding the spinning-tool (roller) into the sheet towards the mandrel to shape the sheet into the geometry of the mandrel. In conventional spinning, sheet thickness is not affected significantly during forming. Different variants of the spinning process have been depicted by Wong et al. [10].

ISF technique refers to clamping the sheet into a fixture and the sheet is not allowed to rotate or move inwards. Incremental forming is accomplished by a forming-tool that traces the tool-path using successive planar or spiral contours. The fundamental difference between ISF and spinning is very simple. In ISF, target geometry is attained by rotating the forming-tool, not the clamped sheet. In addition, sheet thickness is reduced significantly during the ISF process. Differences between ISF and spinning are discussed in Section 2.3.

Most of the researchers have mentioned in the literature that the idea of ISF was patented by Leszak in 1967 (see Appendix A), but this seems to be unsatisfactory if we consider the process definition as described above. In order to find out the origin of the ISF process and first true ISF patent, a timeline of historical developments is discussed in this section. However, Jeswiet et al. [11] presented an extensive study on basic aspects and forming-ability of the ISF process in a wider sense. But no attention was paid to the particular technological developments of the ISF process. Hence, the technological developments of the ISF process have been presented over the years.

Special attention has been paid to the technological developments of the ISF process from the automotive industry due to the fact that demand of industries generally leads to innovative ideas. Many older publications and patents have been taken into account for gaining the information related to the origin of each variant of the process.

Depending on technological developments, the history of the ISF process can be divided into four eras as follow:

1. First era (until 1989)
2. Second era or era of Japanese developments (1990 to 2000)
3. Third era (2001 to 2010)
4. Modern era (2010 onwards)

### 2.2.1 First Era (until 1989)

This era can be considered as the period of earliest history when the idea of ISF was originated. However, patents issued in this era cannot be associated with modern ISF. Modern ISF techniques evolved from spinning processes. In the late 1960s, many patents were issued on different variants of spinning processes which were close to the ISF technique but cannot be regarded as true ISF. Two US patents can be taken as examples, both issued in 1967, the first issued to Leszak, and the second issued to Berghahn of General Electric (see Appendix A for a full list of patents). The Leszak patent [US 3342051] proposed a process for producing cup-type symmetrical shapes in which a roller-type tool presses the sheet material into an elastic medium. The sheet is rotated on the mandrel and vertical displacement is made due to local bending of the sheet material. The Berghahn patent [US 3316745] also proposed a process for producing cup-type symmetrical shapes in which the sheet is clamped in a disc-type arrangement and made to rotate. The roller is made to move towards the center of the disc along a radial line following contour by contour movements.

There is a fundamental difference between both of the proposals. Leszak proposed that the sheet is shaped by a bending action whereas Berghahn described the movement of the roller in three mutually perpendicular directions (X, Y, and Z). However, both the above-mentioned patents describe two different variants of a spinning process; still, the Berghahn process can be considered as an example of an ISF process, and the Leszak process cannot. However, the Berghahn process cannot be considered the origin of the modern ISF process, as there is no evidence in the literature that this patent initiated the developments related to true ISF. The Berghahn process can be regarded as ISF, and the Leszak process cannot. By backtracking the publications on ISF, Mason's [12] work can be considered as the origin of the true ISF process. In his work, he reviewed various sheet forming techniques for batch-type production. He proposed that three coordinates (X, Y, and Z) are required to form a shaped surface. He found that a spherical roller can produce a desired shaped surface with a three axes-controlled apparatus. He also mentioned that a backing material is required for forming the sheet, and his work concentrated more on finding the suitable backing material. However, this assumption (need for a backing material) turned out to be unnecessary with newer developments. At that time, a CNC milling machine was not available commercially, hence he performed experimental work on a lathe machine by rotating the clamped sheet. Later, his work was addressed by his tutor in Kyotoin 1984 [13]. It is believed that this work drew the attention of Japanese researchers, and technological developments shifted towards Japan from the Western world.

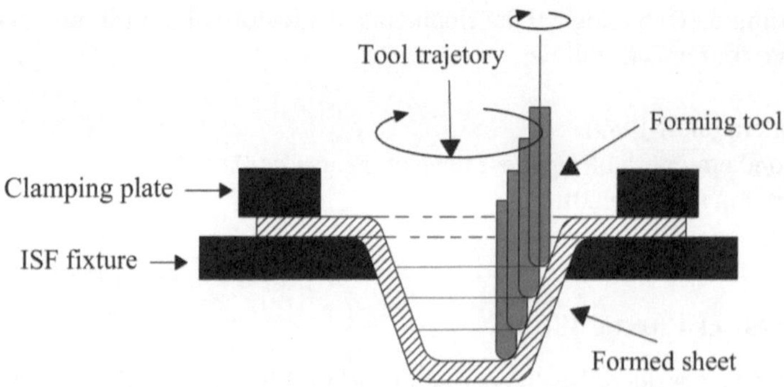


**FIGURE 2.1** Principle of SPIF for a non-axisymmetric shell.

### 2.2.2 Second Era or Era of Japanese Developments (1990 to 2000): Beginning of True ISF Process

In 1989, Iseki [14] and his co-workers became the first Japanese researchers who referred to Mason's work, mentioning that non-symmetrical components can be formed on a CNC milling machine using a 3D tool-path. Iseki explained the schematic of forming the non-axisymmetric shell-type components in his original article [14] which has been adopted as a true ISF or modern ISF process as shown in Figure 2.1. The first set-up was very simple, and the sheet forming was performed on a manually operated three axes machine due to the fact that CNC milling machines available at that time had lower stiffness and accuracy. Furthermore, a CNC milling machine was an expensive machine tool and was not designed for forming operations and was considered unsafe for such operations. Later, Iseki and his co-workers [15–17] reported components of various shapes using computer-controlled axes in their articles. Iseki's work can be regarded as the origin of the true ISF process.

In continuation, Iseki filed a patent on "Incremental Bulging Machine" and "Forming Tool Unit" in 1996 and the same was issued in 1998 [JP 10-180365, P3445988]. In addition, a CNC incremental machine tool was developed by the team [18, 19], which was able to bear a forming load up to 50 kN. They performed significant experimental work on the developed machine tool for forming various materials (including stainless steel and titanium) to test the forming capacity of the former.

Iseki's work paved the way for researchers in the field of ISF in Japan. Kitazawa [20] also explored this process parallel to Iseki's work. He performed experimental tests on a CNC lathe machine (by rotating the sheet) and on a CNC milling machine (by fixing the sheet) in order to form simple shapes. It is noteworthy that the work done by Mason and Kitazawa was in accordance with the principle patented by Berghahn [US 3316745].

However, many research articles on incremental methods of forming were presented at the ICTP 1993 in Beijing, China. But all this was unobserved by the Western world. In 1994, in the special issue of the *Journal of the Japan Society for Technology of Plasticity* published by the Japan Society for Technology of Plasticity, the paper on "Movement of Intelligent Incremental Forming" drew the attention of researchers, and this proved a milestone in the field of incremental forming methods.

### 2.2.3 Process Variation (1993–2000)

It is noteworthy that all above-described variants of incremental forming (except Leszak's patent, which refers to the clamping of the sheet on a mandrel) relate to the clamping of the sheet at its perimeter while a roller or spherical-type tool is allowed to form the sheet incrementally using different control mechanisms. No additional support is provided to the sheet. This method of forming is termed as single-point incremental forming (SPIF) in modern ISF (described in detail in Chapter 3). The tool moves over the surface of the sheet in a planar motion taking a downward step-depth after each planar contour, also moving towards the center of the sheet, as shown in Figure 2.1. No patent has been issued on this type of process variant directly. However, several open publications are available on this process variant.

In 1995, a Japanese patent [JP 07-132329] was issued to Matsubara for a different process variant; nowadays it is known as two-point incremental forming (TPIF). In TPIF, the sheet is supported at its center and its perimeter is deformed in a vertical downward direction by the tool-movement. The stylus or tool moves from inside to outwards in a planar motion. Step-depth is given to the forming-tool contour by contour in the downward direction synchronously with the perimeter of the sheet (described in detail in Chapter 3). This variant was originally presented by Matsubara [21]. This patent [JP 07-132329] has some special features:

- This patent deals with the method as well as apparatus.
- The mid-support to the sheet is a simple bar.
- Blank is passively moved in a vertical downward direction.

Several patents were issued after this patent in Japan. Most of the patents were filed by companies (see Appendix A). In 1997, another patent of Matsushita [JP 09-010855] provided an extension of the original patent of Matsubara [JP 07-132329]. According to this patent, active controls were provided for the movement of the blank. In 1996, Hitachi filed a patent [JP 09-085355] that covers a large range of applications using one or two forming-tools. They showed that both tools (upper and lower) can move in a synchronized manner for the successful forming of curved plates. In the same year, Toyota also filed a patent [JP 10-137858] which was another extension of the original patent of Matsubara [JP 07-132329]. This proposed the forming of complex shapes using an NC milling machine. In 1997, Hitachi filed another patent [JP 10-296345] that claimed the use of several forming roller-tools in SPIF. Toyota was the first one to claim the use of full male support in TPIF in 1997 [JP 10-314855]. The sheet perimeter remained fixed instead of moving in the vertical direction. In 1998, Hitachi filed a patent [JP 11-285741] for using the tool-tip of several rollers. In 1998, Toyota [JP 2000-153313] also claimed to produce wrinkle-free components using an incremental process by using a counter-tool in synchronization with a forming-tool. It was claimed that the thinning of the sheet material was decreased compared to that produced by a single forming-tool. To reduce the sheet-thinning, Honda also claimed an ISF process [JP 2002-102944] with pre-shaping operation in 2000.

In 1999, the most relevant patent was filed by Matsubara and Amino [EP 0970764, US 6216508] as an extension of a 1995 patent of Matsubara [JP 07-132329] on TPIF technique. They patented the TPIF apparatus with a flat top to the central support, which is a replica of the bottom shape of the final components. Moreover, two actuators were provided for the vertical movements of the sheet material.

It is clear that this was the period (1993 to 2000) of process variations as many variations were experimented with in the true ISF process. Simple to complex supports were offered to the sheet material which led the process variation of ISF from the SPIF to the TPIF process. Moreover, many companies showed interest in the process, using it to produce some real products. All these developments were made in Japan only, and several Japanese patents were issued for the technological developments of the ISF process (see Appendix A for patents' information). Till 2000, the ISF process remained unnoticed by the Western world.

### 2.2.4 Third Era (2000 to 2010)

During this period, research work was transferred from the Eastern world (especially Japan) to the Western world (especially Europe). In 1997, a College International pour Recherheen Productique (CIRP), France, meeting was attended by various researchers of the Western world, where the simplicity and flexibility of the ISF technique attracted them. By this time, the availability of CNC milling machines was not a big issue in Western countries. Moreover, the stiffness and accuracy of these machines had become reliable. The patent issued to Matsubara in 1995 became the inspiration for many inventions in the field of ISF. In 2001, a patent [US 6532786] was filed by Luttgeharm (D-J Engineering, Inc.) under the name "Numerically controlled forming method" and the same was issued in 2003. This method was described as a variant of the spinning process in which the workpiece was kept stationary. In this variant, the forming-tool moves over the surface of the sheet, and the sheet is supported by full die. It is noteworthy that the sheet perimeter was not clamped but was allowed to shift downwards and inwards. The sheet was simply supported with the help of some clamping medium (like a screw) at its center. This process was all but identical to the TPIF process with full die. Luttgeharm filed another patent with the same title but with a different claim in 2003 which was issued in 2004 [US 6748780]. Several forming-tools were discussed in this patent. Furthermore, the inventor correlated this process with shear spinning. "*Shear forming*" was said to be the deformation mode of this process.

#### 2.2.4.1 Warm Forming

*Warm forming*: Many other patents have been found in the literature related to the incremental forming of sheets. In 2001, Okada and Kawamura (Hitachi Ltd.) filed a patent [EP 1147832, US 6561002] under the name "Incremental forming method and apparatus for the same" which was issued in 2003. This claimed method was a combination of ISF and the deep-drawing process. In 2003, Okada and his coworkers proposed another patent [EP 1462189, US 6971256] which was the joint effort of Hitachi Ltd. and Amino Corporation. They proposed a heating device to the TPIF set-up and claimed to have reduced spring back effect during forming operation. In 2006, the University of Leuven, Belgium, filed a similar patent [EP 1899089] in which the laser-assisted spinning apparatus was provided with a heating device. A similar system was presented by Klocke and Wehrmeister [22].

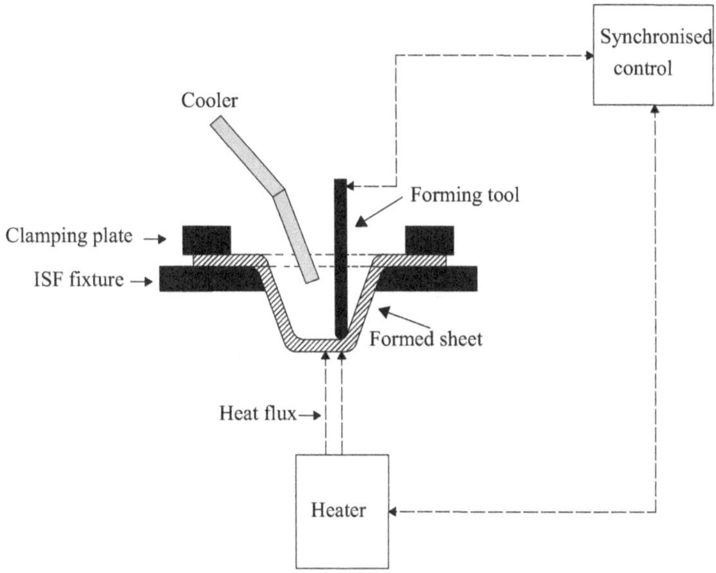


**FIGURE 2.2** ISF with additional heating.

In their arrangement, local heating is accomplished by a heating device that follows the master-tool on the back side of the sheet. Figure 2.2 represents the schematic of the arrangement patented by the University of Leuven, Belgium. Later, Duflouet al. [23] and Mohammadi et al. [24] also worked on a laser-assisted ISF technique. In 2006, another patent request was filed by Callebaut and co-workers [US 0158805, identical to EP 1899089] under the same name (Asymmetric Incremental Sheet Forming System) claiming some extended aspects. A loco-regional heating/cooling system was arranged to follow the forming-tool in order to control the temperature of the forming-zone. It was claimed that the required forming-forces were reduced significantly using a heated ISF process. The formability of the heated sheet was also said to be increased. In addition, the geometrical accuracy of the formed components was claimed to be increased when the sheet material was formed using a heating device.

Later, many studies were introduced for local heating of the sheet during the forming operation. Park et al. [25] were among those who proposed friction as a source for heating the sheet locally. They used a high speed of rotation of the forming-tool to increase friction at the tool–sheet interface. This study claimed that the local temperature reached up to 177°C. Asghar et al. [26] investigated the impact of high current density during the DSIF process on Ti-6Al-4V sheets.

Another variant of TPIF was patented by Okada and co-workers (Hitachi Ltd. in association with Amino Corporation) in 2004 [EP 1462190]. They presented a methodology to form a curved sheet using full die support. In 2004, Amino and his co-worker got a patent [US 0148997] under the name "Shaping method and apparatus of thin metal sheet" in order to accomplish the successful forming of the thin sheets by proposing a process which is a combination of TPIF and stretch-drawing. In this method, a thin sheet is initially formed to intermediate shape by stretch-drawing using a mold-punch. Later, a spherical stylus (tool) is moved over the surface of the sheet, on the other side, in a trajectory defined by the NC environment. Furthermore, this method was modified in 2006 by extending the previous patent. In this issued patent [EP 1731238, US 0272378], a thin sheet was roughly formed using two passes of stretch-drawing followed by incremental forming after each pass.

#### 2.2.4.2 Use of Two Forming-Tools and Robots

ISF had been executed on CNC machines and special machines (as supplied by Amino) so far. It is quite possible to execute the ISF process using heavy-duty industrial robots. The use of robots during the ISF process increases the flexibility of this process. On the other hand, the lower stiffness of these robots limits the forming capacity of the process. However, this problem can be eliminated by using a heating device in order to form hard-to-form materials (magnesium and titanium alloys). The principle of using industrial robots for incremental forming was patented by Tuominen [EP 1560668] in 2004. In this patent, Tuominen described a process (similar to TPIF) in which sheet forming was accomplished using a supporting die, and the CNC milling machine was replaced with an industrial robot. The forming-tool (which was clamped to the robot's wrist) was allowed to move over the surface of the sheet in order to form the target geometry. In addition, Tuominen also proposed a strategy to accomplish the ISF process using two robots. In this strategy, a supporting die was replaced by a moving tool that was controlled by the second robot. This process was considered an extension of the TPIF process. The same process was also reported in Europe by Strano [27]. Jadhav [28] elaborated this process in his thesis and termed this process kinematic ISF. This concept was actually explored by Meier et al. [29] taking two industrial robots into use. In 2008, Wang et al. [30] proposed another variant of this concept. In this variant, both tools were clamped in a single frame. The control for the second tool was eliminated. In 2008, Franzen et al. [31] also executed this process using two forming-tools (one at the top of the sheet, another from the back side of the sheet). The second tool was a cone-shaped block placed on the rotating table. This process was named Dyna -Die and was limited to produce symmetrical shapes only as the sheet was clamped on a rotating table which was directly controlled by a motor. A robot was not used in this (Dyna-Die) variant of ISF. The similar type of variant (using two forming-tools without a robotic set-up) was patented [EP 1977842] by Rodriguez Gutierrez of Fatronik, a Spanish R&D institute. Two forming-tools on the same side of the sheet were used in order to increase the speed of the ISF process by Kwiatkowski et al. [32].

### 2.2.5 Modern Era (2010 to till Now)

In recent times, many strategies of forming components have been combined with the incremental forming by researchers and inventors. The novel technique proposed by Callebaut and co-workers [US 0158805, identical to EP 1899089] for improving the formability and reducing forming-forces paved the way in the field of heat-assisted ISF. It is noteworthy that the forming-force to deform the material and the formability of the material are two critical output factors of the ISF technique. These factors decide the size and design of the forming apparatus significantly. As an instance, greater forming capacity of the forming apparatus would be required when larger forming-forces are needed to deform the sheet material. In addition, more wear would occur in tools and dies (if any) required for the forming operation. A larger forming apparatus would consume a higher amount of energy. All these factors raise the overall cost of the product. Therefore, any technique or process is really appreciable for reducing the manufacturing cost of components that can reduce the required forming-force and can enhance the formability of the sheet material. Hence, it is appreciable to smaller equipment to perform the forming process. Claiming a solution to the aforementioned issues, Roth (Penn State Research Foundation, US) filed a patent request [US 8021501] in 2008. They claimed to introduce a technique of forming components at elevated temperatures in the SPIF process using an electric current. The proposed apparatus was nothing but a simple SPIF set-up which was connected to DC supply. The anode terminal of DC was connected to the forming-tool whereas the cathode terminal of the DC source was connected to the work-table. They tested many metallic materials (6Al-4V titanium alloy, 6061T, 6511 aluminum alloy, 7075 T6 aluminum alloy, C11000 copper alloy, 464 brass alloy) using this technique and showed various relationships amongst flow stress, strain, and current density. Direct current was supplied to the forming-zone. They also claimed to supply and terminate the electric current at any time of forming as a special feature. Forming-force and spring back effects were claimed to be decreased and the formability of the components was said to be increased. Nevertheless, there are some drawbacks of forming at elevated temperatures like reduced lubrication effect, additional set-up of heating source, changed structure of the sheet metal, etc. Later, Fan et al. [33] and Honarpisheh et al. [34] also worked on electric-assisted ISF process and claimed to increase the ductility of the forming-zone of the sheet material.

In 2009, Smith and co-workers requested a patent [US 0226272] combining two methods (milling and forming) which could be performed on a CNC milling machine. They proposed to manufacture the components (which require two operations) in a single set-up by employing a milling-tool for cutting the thin sheets, and a forming-tool for shaping the sheet metals using an SPIF technique. Unfortunately, this invention did not find further continuation by researchers due to many limitations. It is noteworthy that the ISF process is adopted by researchers and inventors due to its key benefit and that is the flexibility. The combination of two kinds of tools limits the freedom of the variation of this novel process.

Johnson and his co-workers (Ford Global Technologies, LLC, US) filed a patent [US 0199742] in 2009 for claiming a variant of the ISF process that uses two forming-tools at a time. This concept was different from previously mentioned work. In this concept, both tools were supposed to move along a predetermined path on a CNC milling-type machine. Both forming-tools were able to form the sheet area independently. Moreover, a heating device could also be attached with both forming agents enabling different amounts of heat-supply at both sides of the sheet. According to another claim of this patent, this concept can be performed using two industrial robots.

An extension of Johnson's work [US 0199742] was filed by Kiridena under the same assignee (Ford Global Technologies, LLC, US) in 2010 which was patented in 2012 [US 8302442]. In this patent, the inventors proposed a feature-based tool-path using two manipulators at a time (one from the upper side of the sheet, the second from the opposite side) to target the desired geometry, taking the squeeze factor of the sheet material into account. The manipulators were allowed to work in a five-axis environment. This technique is being adopted by researchers and termed as double-sided incremental forming (DSIF) (described in detail in Chapter 3).

In continuation, Ren and Xia requested a patent [US 0024034] in 2010 for a method of enhancing the geometrical accuracy of the process patented [US 8302442, which is identical to US 0199742] by the same assignee (Ford Global Technologies, LLC, US). A stiffening feature was implemented for reducing the gap between both the forming-tools.

Ford Global Technologies, LLC, US, increased its interest in the previous patented process [US 0199742] which proposed to offer a new forming process executed using two forming-tools moving over the sheet surface (one on the upper surface, another on the opposite surface of the sheet).In 2010, Kiridena and co-workers of Ford Global Technologies, LLC, US, filed another patent [US 0011915] to claim some different embodiments in the process which was already patented [US 8302442, which is identical to US 0199742]. In this patent, they claimed to produce wrap surfaces (in this case, convex and concave shapes) of the sheet material using the DSIF technology.

Ford Global Technologies, LLC, US, had taken the opportunity to claim different embodiments of the DSIF technique in the field of ISF. In a continuation of their claims, Roth (Penn State Research Foundation, US) filed an international patent request [WO 2012/030835] in 2011 to claim embodiments in which the DSIF process was proposed with electric-assisted technology. This process was nothing but the implementations of the findings of the SPIF technique (which was filed in a patent [US 8021501] in 2008) into the DSIF technique. Forming-force and spring back effects were claimed to be decreased, and the formability of the components was said to be increased. Nevertheless, there are some drawbacks of forming at elevated temperatures like reduced lubrication effect, additional set-up of heat source, changed structure of the sheet metal, etc. Later, Asghar et al. [26] investigated the electric-assisted heating technique during the DSIF process and reported similar outcomes.

However, the DSIF process can be executed using two forming-tools (tools can be mounted on industrial robots or a special ISF machine). The sheet can be squeezed (using squeezing tool-path) by moving the forming-tools towards each other. The main issue for this variant of ISF is that if the thickness prediction of the sheet is inaccurate (for assigning tool-path), this variant of ISF turns into SPIF as only one of the forming-tools would be in touch with the surface of the sheet. To overcome these issues, Cao and Malhotra of Northwestern University, US, proposed a patent request [US 0103177] in 2012. A system and method to incrementally form the sheet material using two forming-tools, moving assemblies and a control unit which they termed as accumulative double-sided incremental forming (ADSIF). One of the forming-tools was made to move in the first direction in order to form the sheet material while the second forming-tool was made to move in the direction opposite to the first one, and vice-versa. They compared the geometric accuracy of the SPIF and DSIF processes with the new invented variant, i.e. ADSIF. The geometric accuracy of the desired components was found to be better using the proposed technique. They also claimed some embodiments related to the tool-path generation, control-unit, and feature-based tool-path strategies. In continuation, Malhotra et al. [35] worked on an ADSIF process for analyzing the impact of wall angle on forming-forces.

In 2014, Ren and Xia of Ford Global Technologies, LLC, US, filed a patent request [US 10010920] which was nothing but the extension of the previous patent [US 0024034]. They claimed to have better geometrical accuracy by reducing the gap between the first and second (which is in the opposite direction) forming-tools for supporting the stiffening feature during the DSIF process.

In 2009, Huskamp and his co-workers (Boeing Company, US) requested a patent [US 0257909] for a method and apparatus able to form sheet material into a useful product requiring less forming-forces during the TPIF process. More particularly, the yield strength of sheet material was made to reduce around the forming-zone by heating during the process. They claimed that the generation of heat could be controlled using a thermal control system.

Another patent [US 9682418] was filed by Young and his co-workers from the same company (Boeing Company, US) in 2009 which was the extension of the previous patent [US 0257909]. However, the claims were different except the TPIF set-up for forming sheet metal components using incremental forming and the same was issued in 2017. A novel lubricating method was proposed in this patent. The lubricant was made to spray over the forming-zone before impinging the forming-tool on the sheet surface. The proposed apparatus comprised of a special kind of forming-tool that was made of a hollow rod having a hemi-spherical end with a texture to supply the lubricant to the hemi-spherical end. A ball-tool was also proposed with the same arrangement for supplying the lubricant on the forming-zone through the forming-tool.

This patent filed in 2009 has not been exploited by researchers due to some limitations, for example it is suitable for low-viscosity forming oils and is not appropriate for highly viscous lubricant (like grease) and solid lubricants (graphite, molybdenum disulfide, etc.).

Nevertheless, the productivity of the ISF technique is one of the major barriers in commercial manufacturing of prototypes and small-volume components due to certain reasons. Loading and un-loading of the sheet into the ISF fixture is one of the major issues which is responsible for increasing the cycle-time of the product. Only one blank is allowed at a time to be shaped in the existing ISF fixtures. In addition, the forming of small size parts by using an existing fixture (if it is of large/standard size) increases the scrap and affects the geometrical accuracy of the formed parts. In order to solve these issues, Kiridena and Xia of Ford Global Technologies, LLC, US, filed a patent [US 0042683] in 2012, proposing a special fixture assembly for ISF set-up that allows sub-divided binder openings of different sizes in order to form different sizes of prototypes in single set-up without un-loading the workpiece. To clamp multiple sheets in the binder, various clamp-supporting segments were employed in the fixture assembly. This invention claimed to eliminate the requirement of reloading the binder after shaping each part. Many components of different sizes can be formed in single set-up which would lead to an increase in the productivity of the process. In addition, various actuators (manually, pneumatically, hydraulically, or electrically) were suggested for clamping the sheet material at its periphery.

In continuation, Nonomura and his co-workers of Toyota Jidosha Kabushiki Kaisha, Japan, filed a patent [US 9522420] in 2013 for improving the dimensional accuracy of the ISF process. They claimed to have improved the dimensional accuracy of the components formed by the ISF technique by stretching the sheet gradually in small steps during the incremental forming by hemi-spherical shaped stylus.

In the ISF technique, various tool-paths are possible for accomplishing the forming operation. These tool-paths are responsible for determining various output parameters of the process like the minimal sheet thickness obtained during the operation. Furthermore, the orientation of the forming-part has a lot to do with the tool-path of this die-less process. In this regard, The Boeing Company, US, filed a patent [US 9676019] in 2013 claiming a system and method for precisely defining the suitable orientation of the forming-part relative to the direction of the forming-tool and a plane that is parallel to the sheet-plane. The applicant claimed to maximize the thinnest segment of the formed component by employing the orientation of the workpiece using the above-mentioned criteria from a plurality of the possible orientations. The performance of the formed components could be enhanced by increasing the thickness of the thinnest portion of the parts.

The dimensional accuracy of the incrementally formed components is largely determined by the springback effect of the sheet material and the stiffness of the forming-tool. Generally, tool-tip and tool-shank of forming-tools are provided tapered interface. This tapered interface leads to the interference between the shank-edge and wall of the components (described in detail in Chapter 3). In addition, the tip-to-shank interface is prone to breakage, being the weakest portion of the forming-tool. These are hindrances to achieving the predetermined dimensional accuracy. In the patent request [US 10144048], Ford Global Technologies, LLC, US, claimed to propose a tool-geometry and arrangement that is able to achieve the desired level of accuracy. The proposed forming-tool arrangement is comprised of a tool-tip, tool-shank and tip-to-shank interface. The diameter of the forming-tip was kept greater than the diameter of the tool-shank so that the interference between the shank-edge and wall of the components can be avoided. The forming-tip was given different shapes (donut-shaped and spherical-shaped). This patent also proposed to employ more than one forming-tool (donut-shaped or spherical-shaped) of different sizes in a single set-up to accomplish multi-pass incremental forming of components.

In 2015, a patent [US 0082498] by Beltran and his co-workers proposed a system and method that is capable of employing the DSIF process economically and accurately. The claimed system included a frame configured to assemblies which are connected to the workpiece, and first and second forming-tools. The claimed system was also capable of introducing heat, by using an electric current, to the sheet material so that the ductility of the sheet material could be increased and the components could be formed with lower forming-forces. The proposed system was equipped with thermal infrared cameras (for monitoring workpiece temperature), load-detecting devices (to avoid overloading of machine-tool), and an optical measurement system for real-time monitoring of the workpiece geometry and tool-displacement. The proposed system, supported by modern equipment, was also claimed to be user-friendly.

Sometimes, a specific thickness and stiffness at certain regions of the component must be attained in order to meet the service requirements. To achieve this, additional material can be adhered/coated to the surface of the sheet. The coating may be of another material. For example, a brass coating can be given to the primary sheet of steel/aluminum. Such a tailored sheet can be employed to form components with different thicknesses and stiffnesses in certain regions of the component. In 2016, Belk and Prakash of "The Boeing Company, US", filed a European patent [EP 3078433] for making sheet metal components from tailored blank using the ISF technique. The inventors also claimed a system that was configured to the ISF machine (robot/CNC milling machine) and was able to apply thermal energy to the forming-zone of the tailored blank in order to control the mechanical properties of the sheet at the deformation-zone.

Another patent [US 0113432] was filed by "The Boeing Company, US", in 2016 that proposed a helical tool-path algorithm for ISF processes. The proposed tool-path had the ability of acquiring different central-axes according to the nature of complexity of desired shapes.

In continuation, Northwestern University, US, filed a patent [US 10162329] in 2017 that claimed a method of generating tool-trajectories for the DSIF method automatically. The proposed technique works on recognizing the features of the component geometry followed by combining the projection map of the desired part.

As studied so far, the components can be formed using various variants of ISF like SPIF, TPIF, and DSIF. The DSIF method has the power to produce high geometric accuracy. Nevertheless, the surface quality of the formed components is still a challenge as the forming-tool produces scar-marks on the surface of the sheet material due to the fact that the forming-tool remains in physical contact with the workpiece during the forming operation. The surface quality of the formed components cannot be compromised. A solution for this issue was disclosed in a patent [US 0030585] issued to Ford Global Technologies, US, in 2019. A method for producing the components using the incremental forming technique was claimed that involves securing the elastomeric material at the outer periphery of the blank and securing an additional layer of sacrificial material to the surface of elastomeric material. The sacrificial material may be a protective film or a low-carbon steel film which can be applied to the workpiece surface with the help of an adhesive layer. The forming-forces exerted by the forming-tool during the deformation of the material are directly applied on the sacrificial material. The parent material does not come in direct contact with the forming-tool and a scar-free smooth surface of the formed components is obtained. This invention was patented for SPIF and DSIF processes.

The above compilation of the history of technological developments of the ISF technique presents an interesting overview of inventions and innovations over the timeline. Many manufacturing companies have shown their interest in this novel die-less technique considering its process flexibility and capabilities. However, ISF is a slow process of forming sheet components and sometimes takes hours to produce specific components, while automotive companies are concerned with quick batch and mass production of components. In addition, universities and research institutes have also focused on this technique for various academic reasons. The technological development of the ISF technique is continuing, and the latest modern technologies are coming in the market in terms of technical and economical competence.

**TABLE 2.1**
**Difference between ISF and Spinning Processes**

| Attribute | Spinning | ISF |
|---|---|---|
| **Sheet edge** | Moves inwards | Clamped at its periphery |
| **Rotating element** | Sheet (usually on mandrel) | Forming-tool (usually in spindle) |
| **Clamping agent for sheet** | Mandrel | ISF fixture |
| **Possible shapes** | Usually symmetrical | Symmetrical and asymmetrical |
| **Wall thickness** | Mostly remains constant | Reduces depending upon process variant |

## 2.3 WORKING PRINCIPLE AND PROCEDURE OF ISF

The principle of the ISF technique is closely related to the spinning process but has some fundamental differences. Both processes are characterized by the local deformation of the sheet performed by a forming-tool in successive steps. In spinning, the sheet (workpiece) is clamped on a revolving mandrel, in order to shape it into the target geometry of the mandrel, the spinning-tool (roller) is fed into the sheet to deform it towards the former (see Figure 1.9). In the spinning process, only axisymmetric parts can be formed and the sheet is pushed inwards by the spinning-tool and sheet thickness is not affected significantly during the forming of the sheet material.

In the ISF technique, the sheet is clamped into a fixture and the sheet is not allowed to rotate or move. Deformation is accomplished by incremental forming by a forming-tool that traces a defined tool-path using successive planar or spiral contours. The fundamental difference between ISF and spinning is that in ISF, target geometry is attained by moving the forming-tool and not the sheet. In addition, sheet thickness is reduced significantly during the ISF process. Table 2.1 presents the fundamental differences between the ISF and spinning processes.

The ISF process is a purely die-less process that uses a simple spherical tool which moves along a tool-path controlled by numerical control (NC)instructions. It is based on deforming the sheet locally layer by layer. This process can be executed on a three-axis CNC milling machine. Figure 2.3 represents the working principle of exercising the ISF process for producing a truncated cone shape. Primarily, the sheet is fixed between the clamping plate and backing plate in the hollow fixture which is directly mounted on the machine table. The forming-tool is fixed in the tool holder which is generally mounted on the spindle of the CNC milling machine or the gripper of the industrial robot. The forming-tool pushes the sheet into a downward direction with a definite step depth. Then it moves in a horizontal plane along the predetermined tool-path. The tool-path is usually generated by CAM packages taking target geometry into account. Figure 2.4 illustrates the various steps involved in the ISF process. The following terminology would be useful for understanding the working procedure of this technique:

- Tool diameter: Diameter of forming-tool used to press the sheet.
- Step size: Amount of material deformed for each contour.

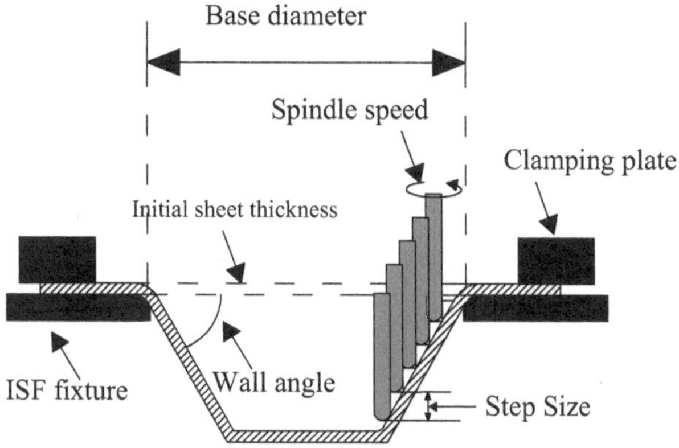


**FIGURE 2.3** Working principle of the ISF technique.

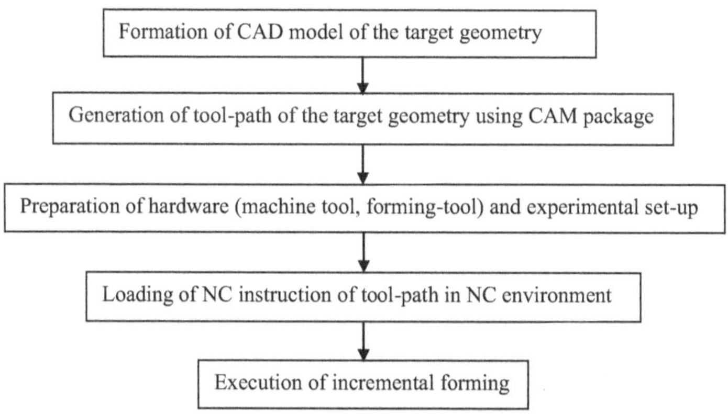


**FIGURE 2.4** Flow-diagram for executing the ISF process.

- Wall angle: Angle between horizontal un-deformed and deformed sheet.
- Sheet thickness: Thickness of sheet before and after deformation.
- Feed rate: Progressive movement of the tool towards the workpiece.
- Spindle speed: Number of tool revolutions at its own axis.

## 2.4 CLASSIFICATION OF ISF PROCESSES

ISF technology can be classified into several types according to its objectives as shown in Figure 2.5. The ISF process can be performed by two different strategies, namely single-stage and multi-stage. In single-stage forming, the blank is clamped in the ISF fixture and then it is deformed with the forming-tool contour by contour in a horizontal plane until the process is completed. A vertical step down is executed after each contour. In multi-stage forming, the components are formed in various stages depending upon the complexity of the target geometry.

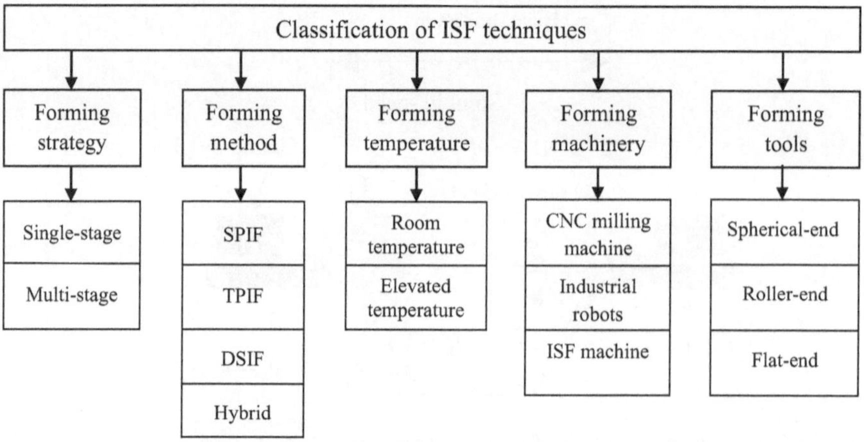


**FIGURE 2.5** Classification of ISF techniques.

According to forming methods, ISF is categorized as SPIF or negative incremental forming, TPIF or positive incremental forming, DSIF and hybrid forming. According to forming temperature, ISF can be executed either at room temperature or at elevated temperatures depending upon the requirement of the process. Forming machinery or hardware to accomplish this novel technique plays a vital role in the productivity and accuracy of the ISF process. Taking this aspect into account, ISF can be performed on either a CNC milling machine or by industrial robots or a special-purpose machine. One other approach to categorizing the ISF technique is its forming-tools. The detailed description of the classification of the ISF technique is presented in Chapter 3.

## 2.5 BENEFITS AND LIMITATIONS OF ISF

Some of the important benefits of the ISF technique include [36–38]:

a) The process can be executed on a CNC milling machine with a negligible cost of set-up.
b) The process is economical for small batch-size production.
c) It is so flexible as compared to other conventional sheet forming processes that the minimum size of a lot can become "one".
d) Despite being used for batch-type production, ISF is very useful for manufacturing parts of old machinery which are otherwise impossible to form due to the unavailability of forming dies.
e) Local deformation at the tool–sheet interface makes the process capable of greater formability than conventional forming processes.
f) Inexistence of the positive or negative die.
g) The forming-forces are smaller than those required by other sheet forming processes like drawing.

However, ISF has been proved an emerging technique of sheet forming due to various advantages as stated above. Nevertheless, the acceptance of this novel technique is restrained as a commercial tool, especially in those applications where geometrical accuracy and forming time are sought as essential requisites. Some major problems encountered during this process are given below [38, 39]:

a) Dimensional accuracy is lower.
b) This technique is not suitable for mass production of components.
c) Vertical walls are difficult to produce.
d) More forming time is required to form a component.

## 2.6 CHALLENGES IN ISF PROCESS

The ISF technique is a still-growing technology that is encountering various challenges. These challenges are barriers to ensuring the commercial acceptance of this die-less technology on the industrial scale. The following are the major challenges of ISF technology:

a) The mechanism of this technology is debatable that needs to be solved by researchers.
b) Interactions between input parameters of this process are quite cumbersome which sometimes leads to material failure.
c) Finite element (FE) simulations of this process are time-consuming and complex.
d) The relationship between forming-force and stress–strain patterns is still questionable.
e) The surface finish and geometrical accuracy of the formed components are poor.
f) The anisotropic behavior and structural integrity of the formed components are inconclusive.
g) The optimization of layout of each variant of the process is not exhaustive yet.
h) The consistent changing of the ISF technologies is also a barrier in defining the common guidelines regarding the process.

The above-mentioned issues and challenges have been encountered while exploring and implementing the process for manufacturing realistic objects for various applications. Moreissues and challenges of the ISF techniques will be discussed in detail in Chapter 7.

## 2.7 APPLICATIONS OF ISF PROCESS

ISF offers a majestic spectrum of applications. In fact, there is hardly any arena of manufacturing that is unaffected by ISF technology. Its versatile nature makes it a powerful potential tool of modern manufacturing for producing medical, automotive, and customized components. This characteristic enables ISF to be a perfect choice for producing numerous customized components of complex shapes. Earlier, some basic shapes (conical, pyramidal, etc.) were seen as the only applications of the ISF technology. Today, the application of ISF is understood to cover a large range of practical aspects. The validity of this interest can be verified by the large number of patents being filed every year. The ISF technology finds applications in the headlight casing of automobiles, automotive service panels, aerospace industry, complex shapes, fuselage parts of airplanes, and customized bio-medical components. The detailed applications in specific industries along with the reasons are explained in Chapter 7.

## 2.8 CONCLUDING REMARKS

ISF technology has evolved into an emerging area of manufacturing during the past three decades. The remarkable progress in the different variants of this die-less technique of forming has inspired researchers and scientists to fabricate various realistic accurate objects of various materials. Various large automotive and commercial companies have paid attention to this die-less technique, taking its process flexibility into account. In addition, universities and researchers have also focused on this technique for various reasons. The principle of accomplishing the ISF technique is quite simple. The advantage of ISF is its process flexibility which makes it suitable for batch-type and prototype production. Today, the application of ISF includes a large range of practical aspects. Nevertheless, full-scale implementation of this process on an industrial scale is still challenging due to the various aforementioned issues and limitations encountered during the process. The technological development of the ISF technique is continuing, and the latest modern technologies are coming in the market in terms of technical and economical competence.

## REFERENCES

1. Pham, Duc, and Stefan S. Dimov. *Rapid Manufacturing: The Technologies and Applications of Rapid Prototyping and Rapid Tooling.* Springer Science & Business Media, 2012.
2. Bagudanch, Isabel, Gabriel Centeno, Carpóforo Vallellano, and Maria Luisa Garcia-Romeu. "Towards the Manufacturing of Near Net Shape Medical Prostheses in Polymeric Sheet by Incremental Sheet Forming." In: Kapil Gupta (ed.), *Near Net Shape Manufacturing Processes*, pp. 1–33. Springer, Cham, 2019.
3. Kumar, Ajay, Vishal Gulati, Parveen Kumar, Vinay Singh, Brijesh Kumar, and Hari Singh. "Parametric Effects on Formability of AA2024-O Aluminum Alloy Sheets in Single Point Incremental Forming." *Journal of Materials Research and Technology* 8, no. 1 (2019): 1461–1469.
4. Kumar, Ajay, and Vishal Gulati. "Experimental Investigations and Optimization of Forming Force in Incremental Sheet Forming." *Sādhanā* 43, no. 10 (2018): 159.
5. Davarpanah, Mohammad Ali, Shalu Bansal, and Rajiv Malhotra. "Influence of Single Point Incremental Forming on Mechanical Properties and Chain Orientation in Thermoplastic Polymers." *Journal of Manufacturing Science and Engineering* 139, no. 2 (2017): 021012.
6. Malhotra, Rajiv, Liang Xue, Ted Belytschko, and Jian Cao. "Mechanics of Fracture in Single Point Incremental Forming." *Journal of Materials Processing Technology* 212, no. 7 (2012): 1573–1590.

7. Kumar, Ajay, and Vishal Gulati. "Experimental Investigation and Optimization of Surface Roughness in Negative Incremental Forming." *Measurement* 131 (2019): 419–430.
8. Kumar, Ajay, Vishal Gulati, and Parveen Kumar. "Effects of Process Parameters on Surface Roughness in Incremental Sheet Forming." *Materials Today: Proceedings* 5, no. 14 (2018): 28026–28032.
9. Blaga, Adrian, Octavian Bologa, Valentin Oleksik, and Radu Breaz. "Influence of Tool Path on Main Strains, Thickness Reduction and Forces in Single Point Incremental Forming Process." *Proceedings in Manufacturing Systems* 6, no. 4 (2011): 191–196.
10. Wong, C.C., T.A. Dean, and J. Lin. "A Review of Spinning, Shear Forming and Flow Forming Processes." *International Journal of Machine Tools and Manufacture* 43, no. 14 (2003): 1419–1435.
11. Jeswiet, J., F. Micari, G. Hirt, A. Bramley, Joost Duflou, and J. Allwood. "Asymmetric Single Point Incremental Forming of Sheet Metal." *CIRP Annals* 54, no. 2 (2005): 88–114.
12. Mason, B. "Sheet Metal Forming for Small Batches." Bachelor Thesis, University of Nottingham, 1978.
13. Mason, B., and E. Appleton. "Sheet Metal Forming for Small Batches Using Sacrificial Tooling." In: *Proceedings of 3rd International Conference on Rotary Metal Working*, 8-10 September Kyoto, Japan 1984.
14. Iseki, H., K. Kato, and S. Sakamoto. "Flexible and Incremental Sheet Metal Forming Using a Spherical Roller." In: *Proceedings of 40th JJCTP*, pp. 41–44, 1989 (in Japanese). Iseki, H., K. Kato, and S. Sakamoto. "Flexible and Incremental Sheet Metal Forming Using a Spherical Roller." *Proceedings of 40th JJCTP* 41 (1989): 44.
15. Iseki, H. "Flexible and Incremental Sheet Metal Bulging Using a Path Controlled Spherical Roller." *Transactions of the Japan Society of Mechanical Engineers C* 58, no. 554 (1992): 3147.
16. Iseki, H. "Flexible and Incremental Sheet Metal Bulging Using a Few Spherical Rollers." *Transactions of the Japan Society of Mechanical Engineers* 59 (1993): 2849.
17. Iseki, H., K. Kato, and S. Sakamoto. "Forming Limit of Flexible and Incremental Sheet Metal Bulging with a Spherical Roller." *Advanced Technical Plastics* 1, no. 1 (1993): 1635–1640.
18. Iseki, H., T. Shioura, and K. Satou. "Practical Development of Flexible and Incremental Sheet Metal Bulging Machine." In: *Proceedings of JSCTP*, pp. 573–576, 1994 (in Japanese).
19. Iseki, H. "Practical Development of Press-Molding Machine with Small Punching Tool." In: *Proceedings of 5th International Conference on Advanced Technology of Plasticity*, Columbus, Ohio, October 7-10, 1996, 2, p. 935, 1996.
20. Kitazawa, K. "Incremental Sheet Metal Stretch-Expanding with CNC Machine Tools." In: *Proceedings of 4th ICTP, Beijing, China, 5 September*, 1993, pp. 1899–1905, 1993.
21. Matsubara, Shigeo. "Incremental Backward Bulge Forming of a Sheet Metal with a Hemispherical Head Tool-A Study of a Numerical Control Forming System II." *Journal of the Japan Society for Technology of Plasticity* 35, no. 406 (1994): 1311–1316.
22. Klocke, F., and T. Wehrmeister. "Laser-Assisted Metal Spinning of Advanced Materials." In: *Proceedings of the Second International WLT-Conference on Lasers in Manufacturing*, Munich, Germany, June 24–26, 2003, pp. 195–200, 2003.
23. Duflou, J.R., Bart Callebaut, Johan Verbert, and Hans De Baerdemaeker. "Improved SPIF Performance Through Dynamic Local Heating." *International Journal of Machine Tools and Manufacture* 48, no. 5 (2008): 543–549.
24. Mohammadi, Amirahmad, LingQin, HansVanhove, MarcSeefeldt, Albert Van Bael, and Joost R. Duflou. "Single Point Incremental Forming of an Aged Al-Cu-Mg Alloy:

Influence of Pre-heat Treatment and Warm Forming." *Journal of Materials Engineering and Performance* 25, no. 6 (2016): 2478–2488.
25. Park, J.G., B.S. You, and Y.S. Kim. "A Parametric Study in Incremental Forming of Magnesium Alloy Sheet." *Transactions of Materials Processing* 17, no. 6 (2008): 412–419.
26. Asghar, Javed, and N. Venkata Reddy. "Importance of Tool Configuration in Incremental Sheet Metal Forming of Difficult to Form Materials Using Electro-Plasticity." *Proceedings of the World Congress on Engineering* 3 (2013): 1734–1738.
27. Strano, M. "Incremental Forming Processes: Current and Potential Applications." *SME Technical Paper No. MF03-114* (2003).
28. Jadhav, S. "Basic Investigations of the Incremental Sheet Metal Forming Process on a CNC Milling Machine. Dr.-Ing." *PhD Dissertation*, University of Dortmund, 2004.
29. Meier, Horst, V. Smukala, O. Dewald, and Jian Zhang. "Two Point Incremental Forming with Two Moving Forming Tools." *Key Engineering Materials* 344 (2007): 599–605.
30. Wang, Yongjun, Ying Huang, Jian Cao, and N. Venkata Reddy. "Experimental Study on a New Method of Double Side Incremental Forming." In: *ASME 2008 International Manufacturing Science and Engineering Conference collocated with the 3rd JSME/ASME International Conference on Materials and Processing*, pp. 601–607. American Society of Mechanical Engineers, 2008.
31. Franzen, V., L. Kwiatkowski, G. Sebastiani, R. Shankar, A.E. Tekkaya, and M. Kleiner. "Dyna-Die: Towards Full Kinematic Incremental Forming." *International Journal of Material Forming* 1, no. 1 (2008): 1163–1166.
32. Kwiatkowski, Lukas, Miroslav Urban, Gerd Sebastiani, and A.E. Tekkaya. "Tooling Concepts to Speed Up Incremental Sheet Forming." *Production Engineering* 4, no. 1 (2010): 57–64.
33. Fan, Guoqiang, L. Gao, G. Hussain, and Zhaoli Wu. "Electric Hot Incremental Forming: A Novel Technique." *International Journal of Machine Tools and Manufacture* 48, no. 15 (2008): 1688–1692.
34. Honarpisheh, M., M.J. Abdolhoseini, and S. Amini. "Experimental and Numerical Investigation of the Hot Incremental Forming of Ti-6Al-4V Sheet Using Electrical Current." *The International Journal of Advanced Manufacturing Technology* 83, no. 9–12 (2016): 2027–2037.
35. Malhotra, Rajiv, Jian Cao, Michael Beltran, Dongkai Xu, James Magargee, Vijitha Kiridena, and Z. Cedric Xia. "Accumulative-DSIF Strategy for Enhancing Process Capabilities in Incremental Forming." *CIRP Annals* 61, no. 1 (2012): 251–254.
36. Micari, F., G. Ambrogio, and L. Filice. "Shape and Dimensional Accuracy in Single Point Incremental Forming: State of the Art and Future Trends." *Journal of Materials Processing Technology* 191, no. 1–3 (2007): 390–395.
37. Pohlak, M., R. Küttner, J. Majak, K. Karjust, and A. Sutt. "Simulation of Incremental Forming of Sheet Metal Products." In: *Proceedings of the Fourth International DAAAM Conference*, 29–30th April 2004, Tallinn, Estonia. 2004.
38. Cerro, Itsasne, Elixabete Maidagan, José Arana, A. Rivero, and P.P. Rodriguez. "Theoretical and Experimental Analysis of the Dieless Incremental Sheet Forming Process." *Journal of Materials Processing Technology* 177, no. 1–3 (2006): 404–408.
39. Verbert, Johan, B. Belkassem, Christophe Henrard, A.M. Habraken, J. Gu, H. Sol, Bert Lauwers, and J.R. Duflou. "Multi-Step Toolpath Approach to Overcome Forming Limitations in Single Point Incremental Forming." *International Journal of Material Forming* 1, no. 1 (2008): 1203–1206.

# 3 Incremental Sheet Forming Process

## 3.1 INTRODUCTION

ISF technology evolved as an emerging arena of manufacturing during the early 1990s. The development of different variants of this die-less technique of forming has inspired researchers and scientists to fabricate various realistic accurate objects of various materials. It has already been discussed in Chapter 2 in detail that ISF has great potential and adaptability to the CNC milling machine and industrial robots. In addition, this die-less variant of forming avoids the need for complex dies and tools, making this technique economical for different applications. The technological development of the ISF technique is continuing, and the latest ISF technologies are coming in the market in terms of technical and economical competence. Nevertheless, this novel variant of forming is yet to be executed at full-scale in manufacturing industries because of several challenges that directly relate to the types of ISF techniques. Classification of ISF technologies is given in Section 2.4.

## 3.2 FORMING STRATEGIES

The forming strategy of ISF implies a number of stages involved in forming the final shape of the components. According to forming strategy, ISF techniques can be classified into two types, viz. single-stage forming and multi-stage forming.

### 3.2.1 Single-Stage Forming

ISF and spinning processes are characterized by the local deformation of the sheet in successive steps performed by a forming-tool. In the conventional spinning process, the sheet edge is pushed inwards by the spinning-tool and the sheet thickness is not affected significantly during forming as the forming is performed by bending only. In shear spinning, the deformation mechanics is stretching instead of bending. The two spinning processes are illustrated in Figure 3.1 for producing a conical shape from a sheet blank. The final thickness of the sheet reduces significantly with the shear spinning technique as the deformation is based on a stretching mechanism [1].

This variation of thickness of the blank is closely dependent on the wall angle of the component that is under consideration (conical shape in this case). This relation can be described by the sine law [1] as shown in Equation 3.1 (see Figure 3.2 also).

$$t_f = t_i \sin(90° - \alpha) \tag{3.1}$$

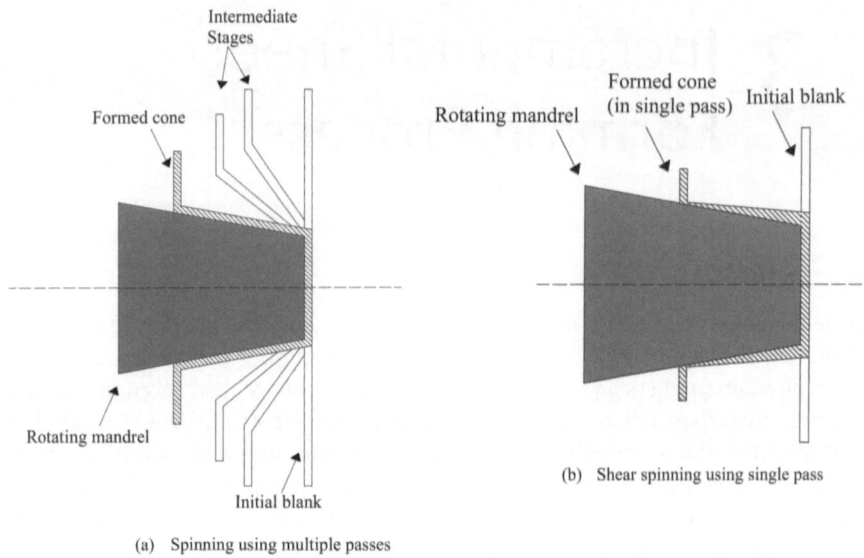


**FIGURE 3.1** Two spinning processes for producing a conical shape: (a) spinning using multiple passes, (b) shear spinning using a single pass.

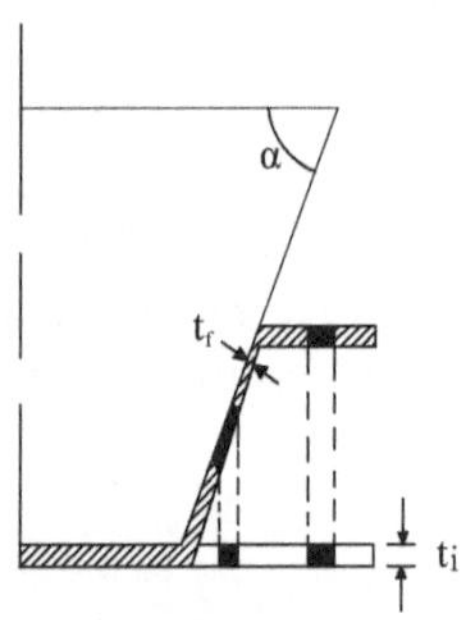


**FIGURE 3.2** Deformation of conical element in shear spinning.

In Equation 3.1, $t_f$ is the final thickness, $t_i$ is the initial thickness of the blank and α is the wall angle. The wall angle is also termed as the drawing angle or forming angle. According to the sine law, an increase in wall angle results in a decrease in the final sheet thickness. Hence, there must exist a limiting wall angle (α) after which failure of the sheet material takes place. It is clear from Equation 3.1 that the sheet thickness would become infinitely small at a 90° wall angle, and that would result in the rupture of the sheet material. The single-stage forming process (Figure 3.3) is also based on local deformation; hence the sine law holds good for the single-stage ISF technique [2–4]. In single-stage forming, the component that is under consideration is formed in a single stage of forming. The forming-tool moves on the horizontal plane of the sheet after each vertical step depth layer by layer and produces the final shape of the component in a single stage [5, 6].

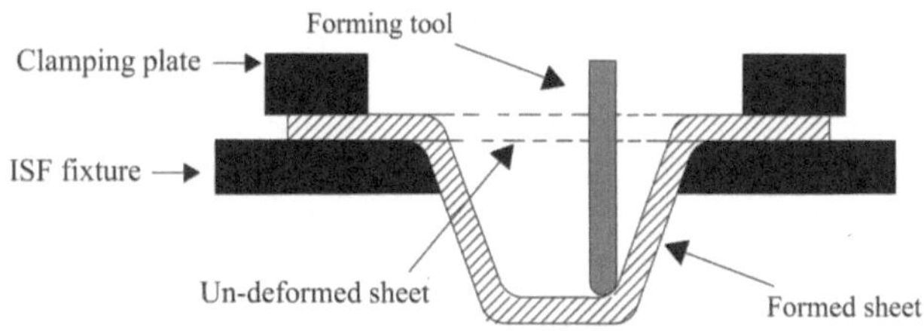


**FIGURE 3.3** Single-stage incremental forming.

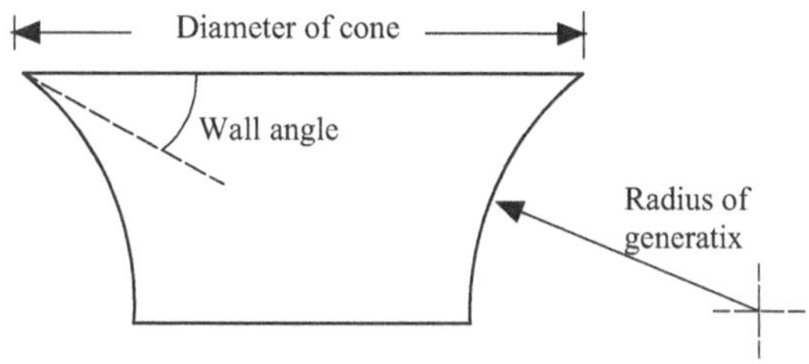


**FIGURE 3.4** Geometrical representation of VWACF.

### 3.2.2 Multi-Stage Forming

It is clear from the above that the wall angle limitation is the common issue in single-stage incremental forming. The limiting wall angle (i.e. maximum wall angle that can be formed without fracture of the sheet material) is normally determined with the help of the cone-forming test. In this test, a varying wall angle conical frustum (VWACF) (Figure 3.4) is formed, keeping other input parameters constant.

In Figure 3.4, the wall angle varies from 0° to 90°. The generatrix of the arc can be of different types (circular, elliptical, parabolic and exponential). It is clear from Equation 3.1 that increasing the wall angle results in decreasing the sheet thickness and increasing the slope of the component with its height due to the sine law. Hence, the component under forming would face fracture at a wall angle less than 90°. In the literature, many researchers [7–10] determined the formability of the sheet material by calculating the limiting wall angle. Figure 3.5 shows some conical frustums

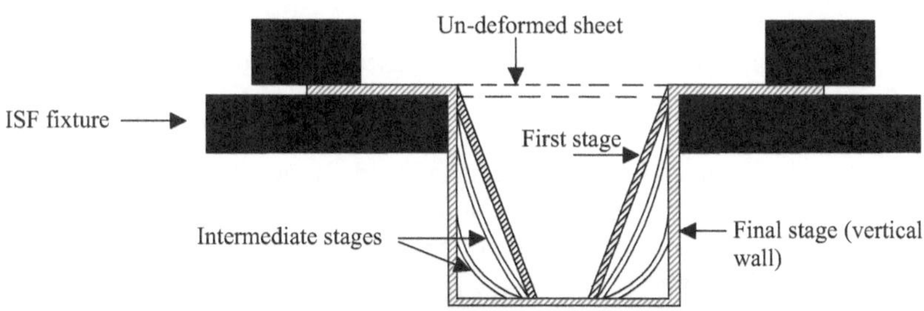


**FIGURE 3.5** Representation of multi-stage incremental forming.

produced by single-stage incremental forming having a varied wall angle. It can be seen that the fracture occurred at an angle less than 90° [11]. Hence, it is impossible to produce a vertical wall using single-stage incremental forming. Now, this becomes the geometrical constraint of this die-less process and needs to be solved. Generally, aluminum and steel alloys can be formed up to a 70° wall angle with favorable forming parameters using single-stage incremental forming [12].

However, the maximum wall angle of the components can be enhanced using the following strategies:

a) *By increasing the initial thickness of the sheet*: The maximum wall angle can be enhanced by increasing the initial thickness of the blank. However, this strategy is not very useful due to the fact that a larger forming-force would be required to form the thicker sheet that in turn would require heavy machinery resulting in increasing cost of the process. Sometimes, product specifications do not allow the thickness of the sheet material to be increased.
b) *By redistributing the material to the forming-zone*: In this strategy, the material from other areas of the sheet can be shifted to the forming-zone. This strategy requires an accurate and complex method that again limits the simplicity and flexibility of this novel forming technique.
c) *By re-orienting the sheet material*: In order to avoid the excessive thinning, because of the increase in wall angle, the sheet can be rotated around the horizontal plane (x–y plane). But again, this approach requires additional help that makes the system complex. Moreover, these strategies do not increase the maximum wall angle significantly. Some researchers have attempted to increase the maximum wall angle by modifying the tool-path and process parameters during single-stage incremental forming [2].
d) *By implementing multi-stage incremental forming*: The difficulty of forming a 90° wall angle in single-stage incremental forming can be eliminated by using multi-stage forming strategies, and several attempts have been observed in the literature for the same [13].

In multi-stage incremental forming, the sheet material is formed into target geometry through several forming stages. Several researchers have claimed to form vertical walls and even undercut shapes using this strategy. Figure 3.5 illustrates the working of multi-stage incremental forming. Generally, the sheet is formed at a 45° to 50° wall angle in the first stage during the forming of the vertical walls of components (like cylindrical cups). Later, wall angle is increased by 10° to 15° in each next stage depending on the geometry requirements of the components.

Primarily, Kitazawa et al. [14] and Kitazawa and Nakane [15] worked on the multi-stage strategy during the SPIF process and formed hemi-ellipsoidal parts using two sequential stages. Later, Young and Jeswiet [16] investigated the thickness distribution on a conical shape and tested the capability of double pass strategy with the sine law. Hirt et al. [17] formed a pyramidal shape having an80°wall angle during the TPIF process. A cylindrical cup with a vertical wall was formed using a five-stage forming strategy by Skjoedt et al. [11]. Similar strategies were confirmed by Skjoedt et al. [13] and Silva et al. [18]. Malhotra et al. [19] proposed in-to-out and out-to-in tool paths for forming a 90° wall angle cylindrical component. Junchao et al. [20] manufactured a tail-light bracket by multi-stage incremental forming.

The strain paths during the multi-stage forming differ from those obtained in single-stage forming [11]. Moreover, the sine law (Equation 3.1) for single-stage forming does not hold good for multi-stage forming. The modified sine law (Equation 3.2) can be adopted in order to incorporate intermediate stages [2].

$$t_f = t_p\left(\sin\alpha_f / \sin\alpha_p\right) \tag{3.2}$$

where $t_f$ represents the final wall thickness, $t_p$ the pre-form wall thickness, $\alpha_f$ the final wall angle and $\alpha_p$ the pre-form wall angle.

## 3.3 FORMING METHODS

According to forming methods, ISF can be classified into the following four types:

a) Single-point incremental forming
b) Two-point incremental forming
c) Double-sided incremental forming
d) Hybrid forming

### 3.3.1 Single-Point Incremental Forming

SPIF is a purely die-less process and basic variant of ISF technology for producing sheet material components. SPIF can be simply employed with the help of a CNC milling machine or industrial robots. In this variant of forming, the blank is simply clamped into a hollow fixture at its perimeter and a simple forming-tool (generally of spherical shape) is moved over the sheet in a horizontal plane with a definite feed and predetermined tool-path [21–23]. The path of the forming-tool is controlled by NC instructions that are normally generated using suitable CAM packages (Del-CAM, UG-NX, etc.). The sheet is pressed by the forming-tool into a vertical downward direction after completing each contour along the horizontal plane. Figure 2.4 represents the SPIF technique. The amount that the sheet is pressed after each contour is termed the step-down size. This parameter (step-down size) is crucial for determining the formability and geometrical accuracy of the components [24]. A detailed discussion of the process parameters is presented in Chapter 5.

This variant of forming is termed SPIF due to the fact that only one point of the sheet is in contact with the forming-tool. This process is also known as "*negative incremental forming*" as the sheet is stretched into the support or fixture. The forming-tool produces local deformation at the contact-zone. The lower surface of the sheet does not come in contact with any partial or full supports. Different stress and strain patterns are observed during the deformation process. The following are the integral features of SPIF:

- Purely die-less process [25].
- Produces concave shapes.
- Produces both axisymmetric and asymmetrical shapes.

- Suitable for low-volume production and prototyping [26].
- Negative incremental forming.
- Only one point of the sheet remains in contact with deforming agents (i.e. forming-tool).
- Greater formability.
- Smaller forming-forces required.
- Quick change in design possible with the same hardware.
- Highly flexible and agile process.

Although SPIF is characterized with many aforementioned key features that increase the flexibility and agility of this process. Nevertheless, some issues have been encountered by researchers that limit the implementation of this die-less process at the commercial level. The following are the problems encountered in the SPIF process:

- Geometrical inaccuracy
- Unwanted sheet-bending [27]
- Greater springback effect [28]

However, researchers have attempted to solve the abovementioned issues during the SPIF process. Geometrical accuracy can be enhanced using adaptive control techniques and modified tool-paths. But these techniques are not enough to solve the problem of dimensional inaccuracy and springback [29]. Unwanted sheet-bending can be avoided by employing a backing-plate to the clamping fixture (detailed information has been given in Chapter 4). Again, this approach would restrict the flexibility of the process as the backing-plate needs to be changed for each target geometry.

### 3.3.2 Two-Point Incremental Forming

As discussed in the above section unwanted bending of the sheet material during forming and springback effects lead to dimensional inaccuracy in the formed components. This reduces the process competency in terms of customers' requirements. The TPIF process solves these problems (springback, unwanted sheet-bending) up to the maximum extent. In this process, a partial or full die is employed to the opposite side of the forming-tool to provide a support to the sheet specimen. The lower surface of the sheet remains in direct contact with the support [27]. This process was originally patented by a Japanese researcher, Matsubara [30]. In TPIF, the sheet is supported at its center only and its perimeter moves in a vertical downward direction along with the tool movement. The stylus moves from the inside outwards in a planar motion. Step depth is executed by the forming-tool contour by contour in the vertical downward direction synchronously with the perimeter of the sheet as shown in Figure 3.6a. This process is termed TPIF due to the fact that two points of the sheet are in contact with forming agents (forming-tool and lower support). This is also known as "*positive incremental forming*" as it produces convex shapes due to the arrangement of the forming apparatus. TPIF can further be divided into two types:

- TPIF with partial die
- TPIF with full die

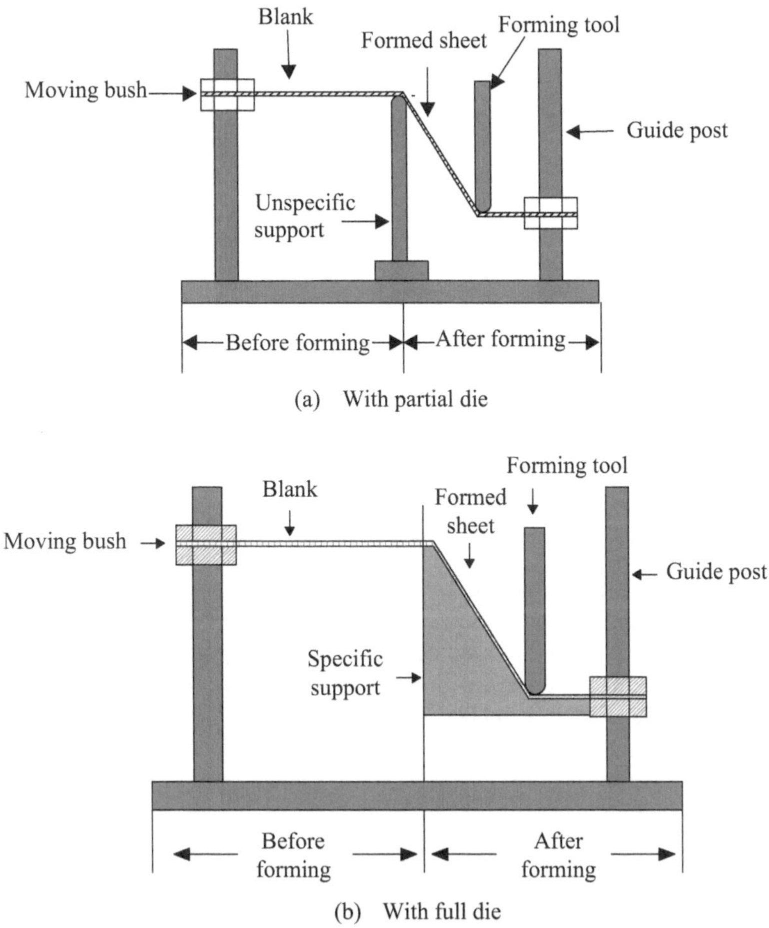


**FIGURE 3.6** The TPIF process: (a) with partial die, (b) with full die.

Figure 3.6a and Figure 3.6b represent the TPIF with partial die and TPIF with full die respectively. The partial die supports only a specific area of the specimen and produces the same effect (enhancing dimensional accuracy) as produced by the backing-plate in SPIF. The same partial die or unspecific support can be easily used to create components of different sizes and similar geometry [31, 32].

On the other hand, TPIF with full die uses different specific support for particular shapes of components, and this variant of ISF is not accepted as a die-less technique. Nevertheless, TPIF with full die considerably enhances the accuracy and precision of the formed components due to the fact that the sheet material is constrained by the forming-punch and specific-die [31, 33, 34].

The die material can be steel, aluminum, plastic, wood, foam or plywood [2]. The following are the integral features of TPIF:

- Produces convex shapes.
- Produces both axisymmetric and asymmetrical shapes.

- Suitable for low-volume production and prototyping.
- Positive incremental forming.
- Two points of the sheet remain in contact with deforming agents (forming-tool and lower support).
- Greater formability.
- Smaller forming-forces required.

TPIF faces the following disadvantages compared to the SPIF process due to the involvement of specific and unspecific dies:

- Quick change in design is not possible due to the involvement of specific dies.
- Poor flexibility.
- Increase in cost.
- Increase in production time.

### 3.3.3 Double-Sided Incremental Forming

The DSIF overcomes the drawbacks of the TPIF process, viz. requirement of specific dies, higher costs and loss of production, without compromising the key benefits of the TPIF process [27]. In this variant of incremental forming, two forming-tools are deployed on both sides of the sheet to produce local deformation as shown in Figure 3.7.

The specific die of TPIF is replaced by a counter tool. This counter tool follows the forming-tool, and the sheet is formed where desired. The main forming-tool acting on the upper surface of the sheet is also known as the master tool, and the counter tool, which follows the master tool and acts on the lower surface of the tool in the opposite direction, is also known as a slave tool. DSIF is a lower-cost method for producing low-volume sheet material components that may not have rotational symmetry. This process was originally proposed by Shima et al. [35] with upper and lower tools moving in a synchronized manner for the forming of curved plates. Later, Murata and Matsuda [36] claimed to produce wrinkle-free components using the DSIF process. Sheet thinning was also found to be decreased in this process.

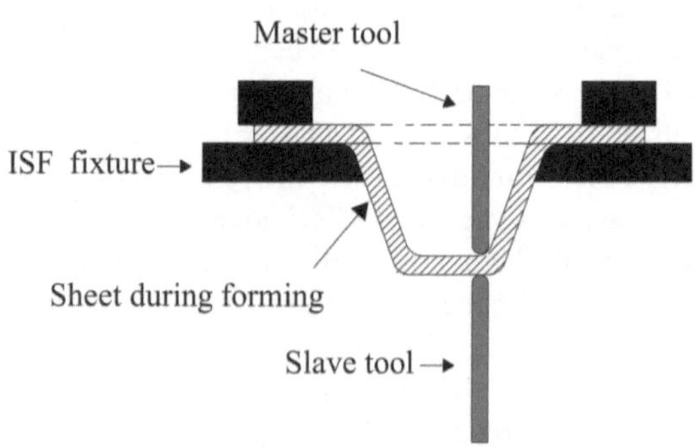


**FIGURE 3.7** The DSIF process.

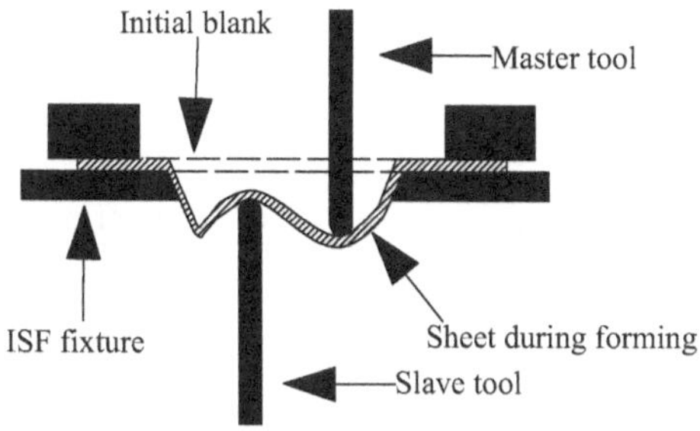


**FIGURE 3.8** DSIF with forming-tools carrying out simultaneous forming.

In DSIF if both forming-tools move synchronously, either a convex or a concave shape is produced, and if both forming-tools are moved independently, a wrap surface (concave and convex shapes on the same sheet) can be produced. In the latter case, both tools are moved along predetermined paths; the forming-tools are able to form the sheet in two areas simultaneously, as shown in Figure 3.8 [37].

The main issue for this variant of ISF is that if the thickness prediction of the sheet is inaccurate (for assigning tool path), DSIF would turn into SPIF, as only one forming-tool would be in touch with the surface of the sheet. ADSIF [28] can overcome this issue. In ADSIF, one of the forming-tools is made to move towards the sheet while the second forming-tool is made to move in the direction opposite to the first one to squeeze the sheet.

DSIF combines the benefits of SPIF and TPIF without the need for specific or unspecific dies. The following are the key features and advantages of DSIF [28, 38, 39]:

- Completely die-less process.
- Can produce both convex and concave shapes.
- Can produce both axisymmetric and asymmetric shapes.
- Suitable for low-volume production and prototyping.
- Can perform positive and negative incremental forming.
- Two points of sheet remain in contact with deforming agents (master tool and slave tool).
- Greater formability and dimensional accuracy as compared to SPIF and TPIF.
- Smaller forming-forces are required.
- No backing-plate required.
- Quick change in design is possible with the same hardware.
- More flexible and agile than the TPIF process.

### 3.3.4 Hybrid Forming

The three types of ISF discussed above, i.e. SPIF, TPIF and DSIF, all follow the identical incremental forming principle of localized deformation of the sheet clamped in a fixture and deformed by forcing the forming-tool with specific movements to achieve the target shape. The localized forming of ISF (SPIF, TPIF and DSIF) results in the physical limitations as follows:

- The localized stretching of the sheet results in localized thinning of the sheet material. Excessive thinning of the sheet material may lead to failure of the material at the forming-zone because the material cannot flow from the outer areas.
- As the deformation is elastic-plastic, significant springback occurs, resulting in deviation from the target shape and dimensional inaccuracy. However, the springback effect is less in TPIF and DSIF as compared to SPIF.
- The repeated movements of the forming-tool over the surface of the sheet material produce residual stresses and strain-hardening that in turn reduces the geometrical accuracy of the components.
- The cycle time of the process is directly proportional to the surface area to be formed and inversely proportional to the step size and feed rate. As a matter of fact, feed rate and step size have limiting values and cannot be increased arbitrarily for reducing the cycle time.

Moreover, the sheet thinning during these variants follows the sine law (Equation 3.1) for single-stage incremental forming. As a result of sheet thinning, a wall angle of the components greater than 70–75° leads to material fracture. However, excessive thinning of the sheet material can be avoided during multi-stage incremental forming. As a matter of fact, various stages are incorporated during multi-stage incremental forming for forming vertical walls in order to avoid excessive thinning at the forming-zone. Hence, the forming time increases multiple times due to the involvement of intermediate stages during multi-stage incremental forming. Hirt et al. [40] produced a square box of $200 \times 200 \times 60$ mm$^3$ which took 15 stages and more than 7 hours to complete forming with multi-stage incremental forming. In addition, the numbers of stages are determined by a "trial-and-error" method in order to produce sound and wrinkle-free components. Hence, multi-stage incremental forming is difficult to apply on complex shapes.

From the assessment of different aspects and variants of ISF (SPIF, TPIF and DSIF), the following limitations increase the physical constraints of ISF:

- For complex shapes, no direct approach solves the problem of excessive sheet-thinning.
- The economic effectiveness of the process is drastically reduced by the long forming-time of these processes.
- Reliable and required tools for robust optimization strategies are not available. Long computer processing time for the simulation of these variants makes the process cumbersome.

Hybrid forming overcomes the above-mentioned limitations. Hybrid forming can be used for producing complex industrial parts economically with improved dimensional accuracy and in less forming-time. Hybrid forming is an integration of stretch-forming and incremental forming. First, a pre-form shape is produced by stretch-forming, and second, the incremental forming is used to complete features

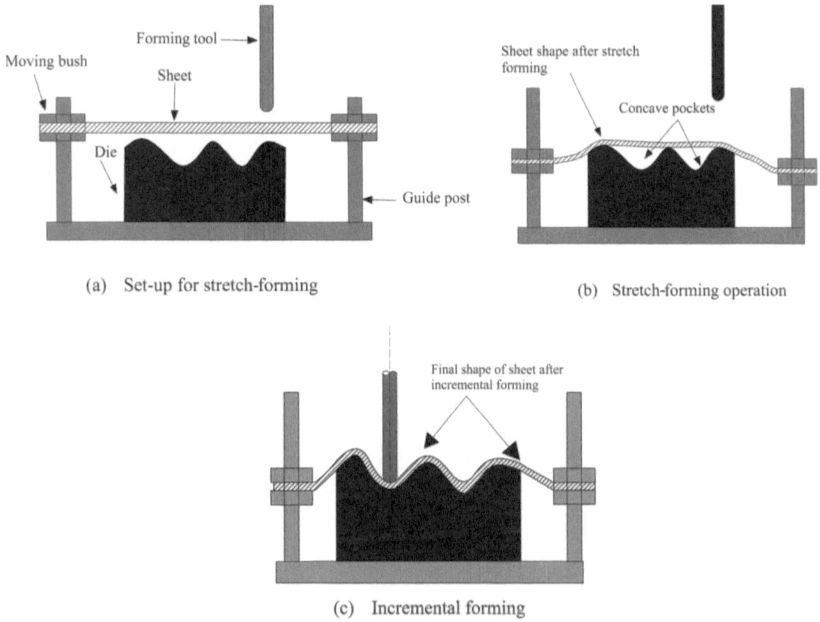


**FIGURE 3.9** Steps involved in hybrid forming technique: (a) set-up for stretch-forming, (b) stretch-forming operation, (c) incremental forming.

like grooves and pockets. The steps of hybrid forming are shown in Figure 3.9 and explained below:

- The blank is clamped in a clamping system (Figure 3.9a).
- The clamped sheet is moved to the die for stretch-forming operation (Figure 3.9b).
- The stretch formed sheet is further formed incrementally by the forming-tool to achieve the target shape (Figure 3.9c).

The pockets or grooves are formed partially during incremental forming, and hence the sheet material does not face strong forming-forces for deformation, leading to the reduction of sheet-thinning [41]. Since the sheet is fixed in a stretched condition after stretch-forming, the elastic effect is avoided, leading to removal of the springback effect making the components dimensionally accurate. Hybrid forming is characterized by the following integral features:

- Cycle time is reduced drastically [42].
- Produces both convex and concave shapes.
- Produces both axisymmetric and asymmetrical shapes.
- Suitable for low-volume production and prototyping.
- Two points of sheet remain in contact with deforming agents (forming-tool and forming die).
- Smaller forming-forces required.

- Thickness reduction of sheet is more homogenous than other variants of ISF [42].
- Tensile forces produced by stretch-forming significantly decrease residual stresses (residual and cyclic stresses are greater in incremental forming due to cyclic movement of the forming-tool).
- Sheet thinning is considerably reduced, in turn increasing the formability of the process.
- The springback effect is negligible, leading to an increase in dimensional accuracy [43].
- The components produced by hybrid forming possess a higher endurance limit than those produced by other variants of ISF [41].

It can be seen from the aforementioned key features and benefits of the hybrid forming technique that it combines the benefits of stretch-forming and incremental forming techniques at the expense of process flexibility and agility. The direct use of specific dies increases the cost of the process and the total production time.

## 3.4 FORMING TEMPERATURE

According to forming temperature, ISF can be executed either by cold forming or by hot forming. Cold forming refers to forming of materials at a temperature lower than the recrystallization temperature of the material under consideration. Most of the time, cold forming is accomplished at room temperature until and unless it is required to raise the temperature of the specimen up to the recrystallization temperature. During hot forming, the temperature of the specimen is maintained higher than the recrystallization temperature using various methods and equipment.

### 3.4.1 Cold Forming

Since the 1990s, ISF has gone through extensive research. Variants of ISF like SPIF, TPIF with partial or full die, DSIF and hybrid incremental forming have been investigated by researchers, and these techniques of incremental forming have been well-developed. During the cold forming of components by ISF techniques, the material is simply deformed layer by layer at room temperature. This localized deformation of the material leads to strain-hardening of the material. In addition, residual stresses are also developed over a span of time as incremental forming of a component requires a long processing time (several minutes and sometimes hours) depending on the geometry of components and forming-parameters (step size, feed rate, etc.). Moreover, the ductility of the material keeps reducing contour after contour due to the strain-hardening of the material. These characteristics of cold forming result in the following issues:

- Formability is decreased.
- Springback is increased.
- Larger forming-forces are required to form the material that has been strain-hardened.

- Higher capacity of the forming machinery is required.
- Higher power is required to accomplish the forming process.

### 3.4.2 Hot Forming

Hot forming is executed at a temperature higher than the recrystallization temperature of the material that is under consideration. The aforementioned issues of cold incremental forming can be solved to a great extent during hot incremental forming techniques. Strain-hardening of the material and propagation of residual stresses during cold incremental forming are the key factors of the limitations that are experienced by investigators and researchers during the cold incremental forming of components. Moreover, some lightweight materials like titanium, magnesium and certain aluminum alloys (AA2024, AA7075) which have shown their significant applications in the biomedical, aerospace and automotive sectors are difficult-to-form materials at room temperature. They exhibit poor formability and higher strain-hardening and springback effects while forming at room temperature.

To overcome these issues and problems, hot or hot forming techniques can be employed with "conventional ISF" (C-ISF) techniques (i.e. ISF techniques executed at room temperature) using various methods and types of equipment. When these hot forming techniques are employed with C-ISF processes, they can be simply termed heat-assisted ISF (HA-ISF) techniques.

The increase in temperature results in increasing ductility of the materials [44–46]. Some difficult-to-form materials like magnesium (AZ31), titanium (Ti grade 2 and Ti grade 5) and aluminum (AA 2024, AA7075, AA5182) alloys have been investigated in the literature and results showed that better formability and dimensional accuracy have been reported at elevated temperatures [47, 48]. In addition, low-carbon steel (DC04) also experienced better formability and reduced springback effects at elevated temperatures.

The heating strategies of HA-ISF can be distinguished into the following two types:

a) Localized heating
b) Global heating
    a) *Localized heating*: In this strategy of hot forming, the sheet is heated locally at the tool–sheet interface. The complete sheet is not heated in this strategy and heat energy of the heat source is saved.
    b) *Global heating*: In this strategy of hot forming, the entire sheet is heated instead of heating a specified portion of the sheet.

Several researchers have investigated the aforementioned strategies of HA-ISF as shown in Table 3.1.

Kim et al. [49] investigated the formability of AZ31 sheets by moving several halogen lamps over the surface of the sheet. Formability was found to increase when the local temperature at the tool–sheet contact zone was raised to 250°C. Palumbo and Brandizzi [50] produced a scaled shell of a car door using static heating by means of electric band heaters. Friction heating was also combined with static heating by employing a high spindle speed (800–1600 RPM). The surface finish of the

**TABLE 3.1**
**Summary of Heating Strategies Investigated in Literature of HA-ISF**

| Heating Type | Heating Device/ Medium | Material of Sheet | Remarks | Reference |
|---|---|---|---|---|
| Local | Halogen lamps | AZ31 | Surrounding area of tool–sheet contact zone also gets heated; hence heating is not purely localized | [49] |
| | Laser | AA5182, Titanium grade 5, and 65Cr2 | The laser source was arranged on the opposite side of forming-tool, i.e. from lower surface of the sheet material. The laser source was dynamically made to follow the movement of the forming-tool | [51] |
| | Laser | | The laser source acted at the same side of the forming-tool and the blank was irradiated at a definite distance from the forming-tool | [52] |
| | Electric | AZ31 | Using an electric source as a heating medium is easy and inexpensive to apply | [53] |
| | Electric | AA2024-T3 and titanium grade 5 | Lower tool diameter and lower wall angle result in burning of the sheet material due to localized heat accumulation because the tool radius and wall angle are inversely proportional to the heat supplied to the sheet | [54] |
| Global | Hot air blowers | AZ31 | Heating of the sheet is not controlled easily | [55, 56] |
| | Heating band | AZ31-O | Heat distribution is not uniform throughout the sheet-area | [57] |
| | Convective heating by hot oil | AZ31B | This set-up is useful for forming only aluminum and magnesium alloys as the hot oil has limits of achieving maximum temperature | [58] |
| Global + local | Heating band + friction using tool rotation | Titanium grade 5 | Temperature distribution was non-uniform due to the position of the heating bands. The entire sheet was heated using a heating band and the localized area was further heated by increasing spindle speed | [50] |

produced components was poor due to the tool-wear that was the result of friction between the tool and the sheet surface. Duflou et al. [51] investigated the forming-forces, formability, springback effects and dimensional accuracy of AA5182, titanium grade 5 and 65Cr2 sheets during hot forming using a laser source as a heating medium. In the laser-assisted ISF, the part was allowed to be heated in a strategically specified area. In addition, the laser source was arranged to follow the forming-tool movement. The forming-forces and springback effects were found to decrease whereas the formability and accuracy of the components were found to increase significantly. Later, Göttmann et al. [52] also used a laser source as a heating medium for employing hot incremental forming. In this set-up, the laser beam was used to irradiate the blank from the same side of the forming-tool.

Fan et al. [53] used an electric source as a heating medium and investigated the formability of AZ31 sheets. Ambrogio et al. [54] also investigated the formability of AA2024-T3 and titanium grade 5 sheets using this heating technique.

Ji and Park [55, 56] used hot air blowers as a heating medium that heated the sheet globally. They investigated the formability of AZ31 sheets at elevated temperatures (higher than 150°C) and found that the formability increased significantly at elevated temperatures. Ambrogio et al. [57] also investigated the formability of AZ31-O sheets during hot forming. The sheet was heated globally using a heating band. The thermocouples were installed with the forming set-up in order to ensure a constant temperature during the forming of sheets. A maximum wall angle of 60° was attained at an elevated temperature (300°C in this case). Galdos et al. [58] used hot oil for heating the sheet globally in order to form AZ31B sheets. The sheet was heated indirectly by the convection method. The forming temperature was maintained at 250°C by means of a temperature-control unit. The maximum wall angle of 60° was attained without failure.

Table 3.1 shows that the local and global heating of sheets has its own benefits and limitations. Local heating requires heating of the forming-area only which directly saves heat energy, making the process efficient and economical. But the heat source is required to move along with the movements of the forming-tool, requiring a control system and complex setup for the movement of the heat source for a safe and successful forming.

In the case of global heating of the sheet material, the control of temperature at the tool–sheet interface is important and incorrect temperatures result in the inaccurate forming. Moreover, excessive heat is required to heat the entire sheet. The heat sources utilized for heating the sheet material for HA-ISF are listed below and described in the following sections.

a) Halogen lamp-assisted ISF
b) Laser-assisted ISF
c) Hot air-assisted ISF
d) Hot oil-assisted ISF
e) Friction-assisted ISF
f) Electric-assisted ISF
g) Induction-assisted ISF
h) Combined heat-assisted ISF

(a) *Halogen lamp-assisted ISF*: In this method of heating, the localized heat is developed using various halogen lamps that move along the forming-tool. This concept was originally used by Kim et al. [49] in ISF on AZ31 sheets as shown in Figure 3.10. However, the heating is not completely localized due to the fact that the larger area surrounding the tool–sheet interface is heated.

(b) *Laser-assisted ISF*: A laser is the source of local heating of the sheet in laser-assisted ISF. This concept of hot forming was introduced by Duflou et al. [51, 59] and Callebaut [60]. They used a heating source of a laser beam on a robotic set-up. Figure 3.11 shows a schematic of the laser-assisted ISF process. A loco-regional heating/cooling system was arranged to follow the forming-tool in order to control the temperature at the forming-zone. It was claimed that the required forming-forces were reduced significantly by using this HA-ISF process. The formability of the heated sheet was also said to be increased. In addition, the geometrical accuracy of the formed components was claimed to be increased. Göttmann et al. [52, 61] used a proportional integral derivative (PID)-based strategy to control the laser heat for attaining the minimum temperature. Mohammadi et al. [62, 63] also investigated laser-assisted ISF and showed that the dimensional accuracy and formability of the components were increased significantly.

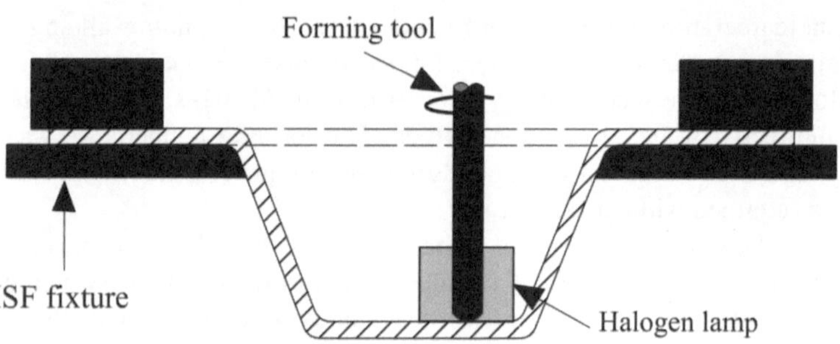


**FIGURE 3.10** Halogen lamp-assisted ISF.

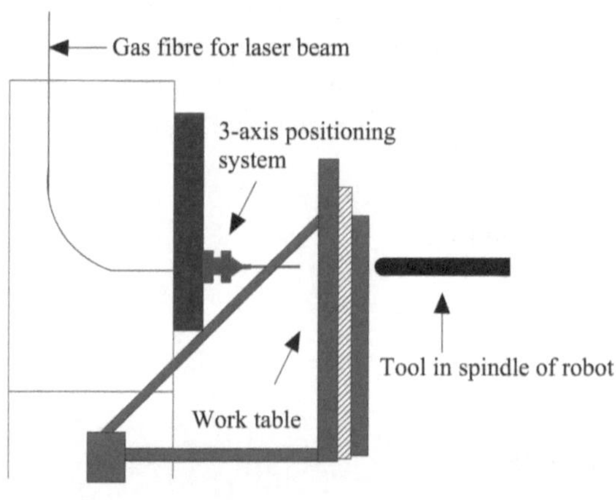


**FIGURE 3.11** Set-up of laser-assisted ISF using robots.

The heating-zone and temperature were well-controlled during this heating technique. Nevertheless, the cost of the experiment set-up is higher, and the process turns out to be uneconomical due to the involvement of several complexities.

(c) *Hot air-assisted ISF*: In this method of heating the sheets, hot air blowers are used as a heating medium. During this method, the sheet is heated globally not locally. This method was originally used by Ji and Park [55, 56] during ISF. They investigated the formability of AZ31 sheets at elevated temperatures (higher than 150°C) and found that the formability of the material increased significantly at elevated temperatures.

(d) *Hot oil-assisted ISF*: In this method of heating, the sheets are globally heated by means of hot oils. The temperature of the heating-oil can simply be controlled by a temperature-control unit. This technique of hot forming was introduced by Galdos et al. [58] during ISF. The forming-sheet was heated indirectly by the convection mode of heat transfer. The forming temperature was maintained at 250°C by means of a temperature-control unit. The maximum wall angle of 60° was attained for AZ31B alloy without failure.

(e) *Friction-assisted ISF*: In this method of heating, the sheet is heated locally by the rotation of the forming-tool during ISF. The friction is generated between the forming-tool and the sheet surface due to the high rotation of the forming-tool. This concept of heating the sheets was proposed by Park et al. [64] during HA-ISF. He used high spindle speeds to increase the friction at the tool–sheet interface. This study claimed that the local temperature at the forming-zone reached up to 177°C. Figure 3.12 shows the schematic view of friction-assisted ISF.

Several researchers [65–71] attempted to investigate the effects of tool-rotation on various responses of ISF like forming-forces, formability, surface roughness, dimensional accuracy, etc. The following are some results from the literature for friction-assisted ISF processes:

- Plastic deformation is accelerated by heating the sheet locally [66, 67].
- At low spindle speeds, the friction between the forming-tool and sheet plays a vital role [68].
- At higher spindle speeds, potential dynamic recrystallization prevails [68].

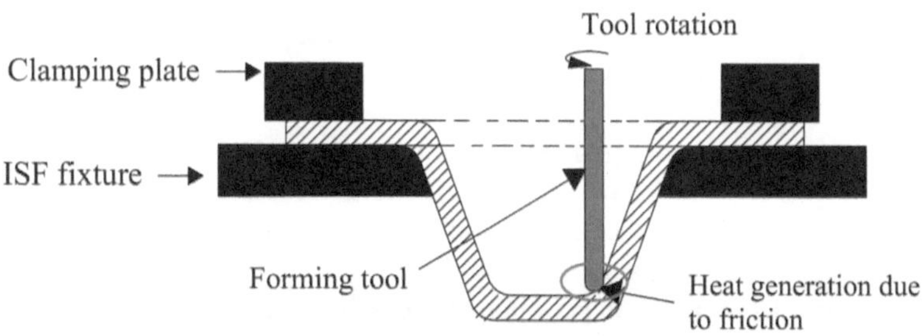


**FIGURE 3.12** Schematic view of friction-assisted ISF.

- This technique is more efficient as compared to other variants of HA-ISF but dependent on component shape [69].
- In the case of polymers, higher spindle speed results in earlier fracture and wrinkling [70].
- The surface micro-hardness, formability and tensile strength of the components are said to be increased by employing high tool rotation [71].
- The formability of the process is increased [24].
- Forming-force is reduced [23].
- Surface quality is poor [5, 22].
- Forming temperature is uncontrollable.
- Excessive tool-wear occurs.

(f) *Electric-assisted ISF*: It is a newer concept than the aforementioned methods for heating the sheet during ISF. In this method of heating, the sheet is heated by an electric current. According to heating strategies (i.e. localized or global), the electric-assisted ISF has been executed by the following two approaches in the literature:

i. Global heating by heating band
ii. Localized heating by resistive heating
i. *Global heating by heating band*: This type of heating provides an alternate approach to heat up the sheet material globally during HA-ISF by using an electric current. In this approach, a heating band is wrapped around the circumference of the ISF fixture. This approach was originally introduced by Ambrogio et al. [57] in HA-ISF. Thermocouples were connected to a PID controller in order to maintain a constant temperature throughout the operation. The forming sheet was heated globally rather than locally. Global heating of the sheet resulted in the reduction of efficiency of heat energy.
ii. *Localized heating by resistive heating*: This aspect of heating provides an alternate approach to heat up the sheet material locally during HA-ISF. The sheet is heated using the principle of resistive heating. According to this principle, the sheet is connected to a cathode terminal of a direct current (DC) supply whereas the forming-tool is connected to the anode terminal. As the electric current is supplied to the sheet through the forming-tool, the localized heating of the sheet occurs. Figure 3.13 represents the schematic of electric-assisted ISF using the principle of resistive heating.

This approach to electric-assisted ISF was introduced by Fan et al. [53] who investigated the formability of AZ31 sheets and found that the maximum forming angle of this material was 64°. Further, Ambrogio et al. [54] tested AA2024-T3, AZ31B-O and titanium grade 5 sheets for formability by this novel approach to forming. Surface quality was found to be poor due to excessive increase of the local temperature. Gottmann et al. [52] and Li et al. [72] attempted to adjust the input current in

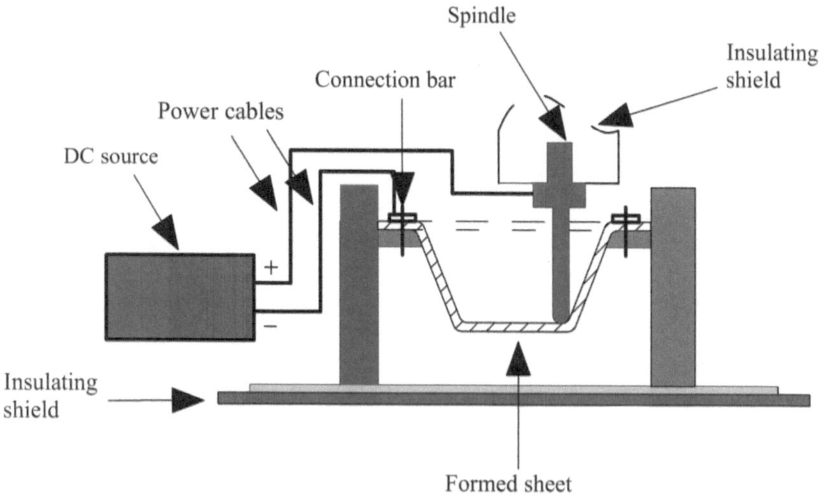


**FIGURE 3.13** Schematic of electric-assisted ISF for local heating.

order to control the forming temperature at the forming-zone. Many other alloys like Ti alloys [73–76] and Mg alloys [77–79] were also tested for electric-assisted ISF. Ti alloys need better control of forming temperature and are difficult to form. Nevertheless, electric-assisted ISF processes for local heating experience some key issues such as the following:

- Inhomogeneous distribution of localized temperature
- Burning of sheets and tool-tip due to high temperature generation at the forming-zone
- Complexity in lubrication system
- Reduction in flexibility of the process

Li et al. [80] introduced an integral electric-assisted ISF process to provide a solution to the aforementioned problems encountered during electric-assisted ISF. But the proposed process also faced some other issues like control of electrode heat, monitoring of temperature and sudden thermal expansion and contraction.

In electric-assisted ISF processes, the flow stresses are reduced significantly because of the *Joule heating effect* and *electro-plastic effect*. According to these effects, extra flow stresses and springback effects are reduced significantly whereas the formability and dimensional accuracy of the components are increased. Due to *Joule heating* (also known as resistive heating), the energy of the electric current is transferred into heat when it flows through a resistance [81, 82]. In addition, the electric current leads to the propagation of the dislocations of atoms in the material, and this is in accordance with the *electro-plastic effects* [83, 84].

Furthermore, it is quite important to describe the value of the "*current threshold density*". It is nothing but the critical density of current below which the effects of electric-assisted ISF are very small. Some researchers [85, 86] have measured it for certain materials and proposed various empirical formulae for estimating this threshold value for several alloys.

In addition, heat supplied to the sheet is inversely proportional to the radius of the tool and the wall angle [54]. Hence, a smaller tool diameter and shallower wall angle can result in burning out of the sheet material at the tool–sheet interface due to the heat accumulation. The selection of suitable lubricants is also a key issue in the electric-assisted ISF process. Some researchers [87, 88] have found that $MoS_2$ powder-coated ceramic results in a better surface quality during this forming technique.

(g) *Induction-assisted ISF*: In this method of heating, the sheet is heated locally at the tool–sheet interface by inductors. This is a simple method to produce hot forming. In this process, the forming-tool moves over the sheet to form the sheet incrementally whereas the inductor provides heat to the forming-zone from the back side of sheet. The heat source (inductor in this case) follows the movement of the forming-tool at the opposite side of the forming-tool (i.e. from the back side of the sheet) in order to produce an elevated temperature at the tool–sheet interface. This method of hot forming was originally introduced by Al-Obaidi et al. [89] in HA-ISF. The set-up arm consists of two sliding fixtures and a slid jig as shown in Figure 3.14. The arm is directly connected to the spindle (by connecting rings) and the induction machine head. Slide jigs and sliding fixtures are so arranged that the inductor follows the movement of the forming-tool during the forming operation to produce localized heating at the forming-zone. The proposed experimental set-up was able to achieve an elevated temperature of more than 850°C. Later, this technique of hot forming was also investigated by Al-Obaidi et al. [90] and Ambrogio et al. [91]. The following conclusions were made during these studies:
- The formability of advanced high-strength steels, DP 980 steel and 22MnB5 steel was increased significantly.
- Forming-forces, residual stresses and springback effects were reduced.
- Dimensional accuracy was increased.

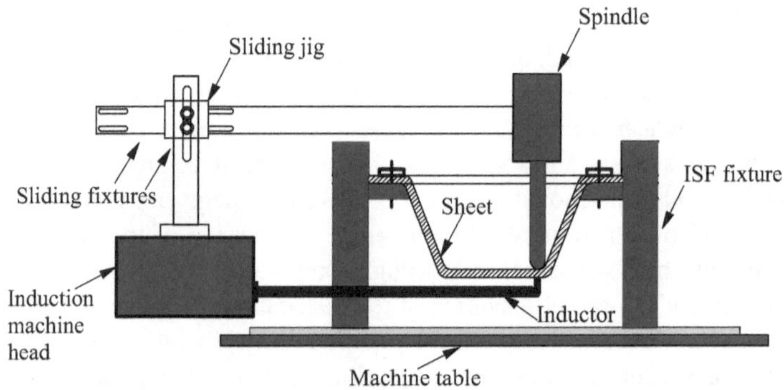


**FIGURE 3.14** Schematic view of induction-assisted ISF.

(h) *Combined heat-assisted ISF*: Combined heat-assisted ISF methods are employed to attain the combined benefits of two different HA-ISF techniques. For example, the electric and friction-generating sources can be used together to employ hot incremental forming for customized components. In addition, the sheet can also be heated locally at the tool–sheet interface after heating globally. Palumbo and Brandizzi [50] introduced a combined HA-ISF technique in which the heating band and friction (produced by tool rotation) were taken into account for accomplishing hot forming. Primarily, the sheet was heated globally using a heating band (as described in electric-assisted ISF) up to a temperature of 150°C. Later, the forming-zone was further heated locally beyond the initial temperature (150°C in this case) by friction (as described in friction-assisted ISF). The temperature at the forming-zone was further increased up to approximately 400°C by employing a spindle speed of 1600 RPM. The following conclusions can be made from the literature:

- The formability and dimensional accuracy of the components is increased.
- Less forming-force is required to produce deformation.
- Tool wear occurs due to the friction between the tool-tip and the surface of the sheet.
- Tool wear is inversely proportional to the surface finish of components.
- Temperature distribution is non-uniform during global heating due to the position of the heating band.

Table 3.2 shows the summary of various HA-ISF methods investigated in the literature. However, various benefits of different hot forming techniques have already been discussed. The mechanism of these effects is still debatable and cumbersome due to the involvement of various complexities [92]. The following are some common issues that are encountered during hot incremental forming techniques:

- Accurate measurement and control of temperature is a difficult task.
- Installing of the temperature sensor is quite a complex task due to the movement of the heating source.
- The thermo-physical properties of the components also limit the accuracy of the non-contact temperature measurement devices.
- Surface emissivity also influences the measurement of temperature and should be measured properly.

However, it is possible to determine and predict the mechanisms of various effects encountered during hot incremental forming by using well-optimized FE simulation [67]. Moreover, simulation of hot incremental forming is quite complex as compared to simulation of incremental forming at room temperature due to the following reasons:

- The thermal stresses should be coupled with the source of heating depending upon the method of heating.

**TABLE 3.2**
**Summary of Various HA-ISF Methods Investigated in Literature**

| Heating Method/ Source | Material Tested | Advantage/s | Limitation/s | Mode of Heat Transfer | Reference |
|---|---|---|---|---|---|
| Halogen lamps | AZ31 | Hardware cost is lower. Simple heating set-up | Heating is not completely localized | Conduction | [49] |
| Laser beam | AA5182, 65Cr2, TiAl6V4, Ti grade 2, AA2024-T3 | Heating-area and temperature range is well-controlled | Experimental cost of set-up is quite high | Radiation | [51, 52, 59, 60–63] |
| Hot air blowers | | Hardware cost is lower. Simple heating set-up | Heating of the sheet is not controlled easily | Convection | [55, 56] |
| Hot oil | AZ31B | Hardware cost is lower. Global heating is employed but the temperature is well-maintained at forming-zone | This set-up is useful for forming only aluminum and magnesium alloys as the hot oil has limits of achieving maximum temperature | Convection | [58] |
| Friction using tool rotation | AZ31, AZ61, AZ80, AA2024-O, AA5052-H34, AA1050-O, AA7075-O, petroleum-based PVC, bio-based PLA, Ti6Al4V | Easy and economical to employ the process. Heating is completely localized | Forming temperature is uncontrollable. Surface quality is poor. Excessive tool wear occurs | Friction | [64–71] |
| Electric current | AZ31; AA1050; AA1060; AA2024-T3; AA5055; AA6061-T6; Ti6Al4V; CP-K 60/78 steel; TC4 Ti; low-carbon steel DC01 | Heating is completely localized. Flow stresses and forming-forces are reduced | Tool wear occurs. Poor surface finish. Lubrication is complex | Electric | [52–54, 57, 72–80, 87, 88] |
| Inductor | DC04 low-carbon steel; DP980 steel; 22MnB5 steel; HCT980C steel; DP1000 steel; Ti6Al4V | Heating is completely localized. Can produce an elevated temperature up to 850°C. Lower hardware cost | Synchronization of forming-tool and inductor is a little complex | Induction | [89–91] |
| Electric current + friction | Ti6Al4V | Lower hardware cost | Temperature distribution is non-uniform. Poor surface finish. Tool wear occurs | Electric + friction | [50] |

- Thermal stresses and strains should be measured during FE analysis.
- Strain rate and effects of anisotropy are different in hot forming and should be measured during FE analysis.
- FE models should be able to process the properties which are temperature-dependent.

## 3.5 CONCLUDING REMARKS

ISF technologies are being witnessed as an emerging area of manufacturing during the past three decades. Various manufacturing applications of prototyping and batch-type production can be satisfied using this emerging technology. Remarkable progress in the different variants of this die-less technique of forming has inspired researchers and scientists to fabricate various realistic accurate objects of various materials. ISF techniques have various benefits and features over conventional forming techniques as described in the above sections. The various issues encountered during cold incremental forming have been solved by researchers to a great extent by employing hot incremental forming techniques. However, accurate measurement and control of temperature have been common issues in hot ISF and are yet to be solved. The technological development of the ISF technique is still continuing, and the latest ISF technologies are coming in the market in terms of technical and economical competence. Nevertheless, this novel variant of forming is yet to be executed at full-scale in manufacturing industries because of several challenges that directly relate to the types of ISF techniques.

## REFERENCES

1. Hagan, E., and J. Jeswiet. "A Review of Conventional and Modern Single-Point Sheet Metal Forming Methods." *Proceedings of the Institution of Mechanical Engineers, Part B: Journal of Engineering Manufacture* 217, no. 2 (2003): 213–225.
2. Jeswiet, J., F. Micari, G. Hirt, A. Bramley, Joost Duflou, and J. Allwood. "Asymmetric Single Point Incremental Forming of Sheet Metal." *CIRP Annals* 54, no. 2 (2005): 88–114.
3. Wong, C.C., T.A. Dean, and J. Lin. "A Review of Spinning, Shear Forming and Flow Forming Processes." *International Journal of Machine Tools and Manufacture* 43, no. 14 (2003): 1419–1435.
4. Hrairi, Meftah, and Salah B.M. Echrif. "Process Simulation and Quality Evaluation in Incremental Sheet Forming." *IIUM Engineering Journal* 12, no. 3 (2011).
5. Kumar, Ajay, Vishal Gulati, and Parveen Kumar. "Investigation of Surface Roughness in Incremental Sheet Forming." *Procedia Computer Science* 133 (2018): 1014–1020.
6. Kumar, Ajay, Vishal Gulati, and Parveen Kumar. "Investigation of Process Variables on Forming Forces in Incremental Sheet Forming." *International Journal of Engineering and Technology* 10 (2018): 680–684.
7. Lehtinen, Pekka, Tapio Väisänen, and Mika Salmi. "The Effect of Local Heating by Laser Irradiation for Aluminum, Deep Drawing Steel and Copper Sheets in Incremental Sheet Forming." *Physics Procedia* 78 (2015): 312–319.
8. Suresh, Kurra, SreeDivya Bagade, and Srinivasa Prakash Regalla. "Deformation Behavior of Extra Deep Drawing Steel in Single-Point Incremental Forming." *Materials and Manufacturing Processes* 30, no. 10 (2015): 1202–1209.

9. Kurra, Suresh, and Srinivasa Prakash Regalla. "Experimental and Numerical Studies on Formability of Extra-Deep Drawing Steel in Incremental Sheet Metal Forming." *Journal of Materials Research and Technology* 3, no. 2 (2014): 158–171.
10. Li, Yanle, Zhaobing Liu, W.J.T. Daniel, and P.A. Meehan. "Simulation and Experimental Observations of Effect of Different Contact Interfaces on the Incremental Sheet Forming Process." *Materials and Manufacturing Processes* 29, no. 2 (2014): 121–128.
11. Skjødt, Martin, Niels Bay, Benny Endelt, and G. Ingarao. "Multi Stage Strategies for Single Point Incremental Forming of a Cup." *International Journal of Material Forming* 1, no. 1 (2008): 1199–1202.
12. Duflou, Joost R., Alexander Szekeres, and Paul Vanherck. "Force Measurements for Single Point Incremental Forming: An Experimental Study." *Advanced Materials Research* 6 (2005): 441–448.
13. Silva, M.B., M. Skjoedt, N. Bay, and P.A.F. Martins. "Formability in Multistage Single Point Incremental Forming." In: *7th EUROMECH Solid Mechanics Conference*, Lisbon, Portugal. September 7–11, 2009.
14. Kitazawa, Kimiyoshi, Akihiko Wakabayashi, Kazuya Murata, and Koro Yaejima. "Metal-Flow Phenomena in Computerized Numerically Controlled Incremental Stretch-Expanding of Aluminum Sheets." *Japan Institute of Light Metals, Journal* 46, no. 2 (1996): 65–70.
15. Kitazawa, K., and M. Nakane. "Hemi-Ellipsoidal Stretch-Expanding of Aluminum Sheet by CNC Incremental Forming Process with Two Path Method." *Japan Institute of Light Metals* 47 (1997): 440–445.
16. Young, D., and J. Jeswiet. "Wall Thickness Variations in Single-Point Incremental Forming." *Proceedings of the Institution of Mechanical Engineers, Part B: Journal of Engineering Manufacture* 218, no. 11 (2004): 1453–1459.
17. Hirt, G., J. Ames, M. Bambach, and R. Kopp. "Forming Strategies and Process Modelling for CNC Incremental Sheet Forming." *CIRP Annals* 53, no. 1 (2004): 203–206.
18. Silva, M. Beatriz, Martin Skjødt, Niels Bay, and Paulo A.F. Martins. "Multistage Single Point Incremental Forming." In: *CMNE 2011 Coimbra, Portugal, January 01.*
19. Malhotra, Rajiv, A. Bhattacharya, A. Kumar, N.V. Reddy, and Jian Cao. "A New Methodology for Multi-Pass Single Point Incremental Forming with Mixed Toolpaths." *CIRP Annals* 60, no. 1 (2011): 323–326.
20. Junchao, Li, Shen Junjian, and Wang Bin. "A Multipass Incremental Sheet Forming Strategy of a Car Taillight Bracket." *The International Journal of Advanced Manufacturing Technology* 69, no. 9–12 (2013): 2229–2236.
21. Kumar, Ajay, Vishal Gulati, and Parveen Kumar. "Experimental Investigation of Forming Forces in Single Point Incremental Forming." In: Shanker Kripa, Shankar Ravi, & Sindhwani Rahul (eds.), *Advances in Industrial and Production Engineering*, pp. 423–430. Springer, Singapore, 2019.
22. Kumar, Ajay, and Vishal Gulati. "Experimental Investigation and Optimization of Surface Roughness in Negative Incremental Forming." *Measurement* 131 (2019): 419–430.
23. Kumar, Ajay, and Vishal Gulati. "Experimental Investigations and Optimization of Forming Force in Incremental Sheet Forming." *Sādhanā* 43, no. 10 (2018): 159.
24. Kumar, Ajay, Vishal Gulati, Parveen Kumar, Vinay Singh, Brijesh Kumar, and Hari Singh. "Parametric Effects on Formability of AA2024-O Aluminum Alloy Sheets in Single Point Incremental Forming." *Journal of Materials Research and Technology* 8, no. 1 (2019): 1461–1469.
25. Kumar, Ajay, Vishal Gulati, and Parveen Kumar. "Effects of Process Parameters on Surface Roughness in Incremental Sheet Forming." *Materials Today: Proceedings* 5, no. 14 (2018): 28026–28032.

26. Gulati, Vishal, Kumar, A. "Investigation of Some Process Parameters on Forming Force in Single Point Incremental Forming." *International Research Journal of Engineering and Technology* 4, no. 4 (2018): 784–791.
27. Beltran, Michael, Jian Cao, and John T. Roth. "System and Method for Incremental Forming." U.S. Patent 9,221,091, issued December 29, 2015.
28. Cao, Jian, and Rajiv Malhotra. "System and Method for Accumulative Double Sided Incremental Forming." U.S. Patent 9,168,580, issued October 27, 2015.
29. Katajarinne, Tuomas. "On the Behaviour of the Process and Material in Incremental Sheet Forming." DOCTORAL DISSERTATIONS, Aalto University, School of Chemical Technology, 27 June 2014 (2014).
30. Shigeo Matsubara. "Method and Device for Forming Metal Plate." Japanese Patent JP 07-132329, 1995.
31. Jadhav, Sanjay. *Basic Investigations of the Incremental Sheet Metal Forming Process on a CNC Milling Machine*. Shaker, 2004.
32. Jeswiet, J. "Incremental single point forming." *Proc. of the ASME Manuf. Eng. Division* 11 (2000): 509–514.
33. Hirt, G., J. Ames, and M. Bambach. "Basic Investigation into the Characteristics of Dies and Support Tools Used in CNC-Incremental Sheet Forming." In: *Proceedings of the IDDRG Conference*, Porto, Portugal, June 19–21, 2006, pp. 341–348.
34. Matsubara, S. "A Computer Numerically Controlled Dieless Incremental Forming of a Sheet Metal." *Proceedings of the Institution of Mechanical Engineers, Part B: Journal of Engineering Manufacture* 215, no. 7 (2001): 959–966.
35. Shima, Akio, Takenao Yoshikawa, Keiichi Nakamura, Yoichi Sudo, and Shoji Suzuki. "Formation of Successively Expanding Metallic Plate and Apparatus Therefor." Japanese Patent JP09-085355, 1997.
36. Atsunobu, Murata, and Fuminori Matsuda. "Incremental Pressing and Forming Device." Japanese Patent JP 2000-153313, 2000.
37. Johnson, Carl Frederick, Vijitha Senaka Kiridena, Feng Ren, and Zhiyong Cedric Xia. "System and Method for Incrementally Forming a Workpiece." U.S. Patent 8,322,176, issued December 4, 2012.
38. Kiridena, Vijitha Senaka, and Zhiyong Cedric Xia. "Method of Incrementally Forming a Workpiece." U.S. Patent 8,302,442, issued November 6, 2012.
39. Ren, Feng, and Zhiyong Cedric Xia. "Method to Improve Geometrical Accuracy of an Incrementally Formed Workpiece." U.S. Patent 8,783,078, issued July 22, 2014.
40. Hirt, Gerhard, Jochen Ames, and Markus Bambach. "A New Forming Strategy to Realise Parts Designed for Deep-Drawing by Incremental CNC Sheet Forming." *Steel Research International* 76, no. 2–3 (2005): 160–166.
41. Choi, Hyunwoo, and Cheolsoo Lee. "A Mathematical Model to Predict Thickness Distribution and Formability of Incremental Forming Combined with Stretch Forming." *Robotics and Computer-Integrated Manufacturing* 55 (2019): 164–172.
42. Araghi, B. Taleb, G.L. Manco, M. Bambach, and G. Hirt. "Investigation into a New Hybrid Forming Process: Incremental Sheet Forming Combined with Stretch Forming." *CIRP Annals* 58, no. 1 (2009): 225–228.
43. Araghi, B. Taleb, Alexander Göttmann, Markus Bambach, Gerhard Hirt, Georg Bergweiler, Jö Diettrich, Marius Steiners, and Alireza Saeed-Akbari. "Review on the Development of a Hybrid Incremental Sheet Forming System for Small Batch Sizes and Individualized Production." *Production Engineering* 5, no. 4 (2011): 393–404.
44. Donachie, Matthew J. *Titanium: A Technical Guide*. ASM International, 2000.
45. Avedesian, Michael M., and Hugh Baker, eds. *ASM Specialty Handbook: Magnesium and Magnesium Alloys*. ASM International, 1999.
46. Welsch, Gerhard, Rodney Boyer, and E.W. Collings, eds. *Materials Properties Handbook: Titanium Alloys*. ASM International, 1993.

47. Emmens, W.C., Gerd Sebastiani, and Antonius H. van den Boogaard. "The Technology of Incremental Sheet Forming—A Brief Review of the History." *Journal of Materials Processing Technology* 210, no. 8 (2010): 981–997.
48. Shi, Xiaofan, L. Gao, Hosein Khalatbari, Yan Xu, Hui Wang, and Ling Ling Jin. "Electric Hot Incremental Forming of Low Carbon Steel Sheet: Accuracy Improvement." *The International Journal of Advanced Manufacturing Technology* 68, no. 1–4 (2013): 241–247.
49. Kim, S.W., Y.S. Lee, S.H. Kang, and J.H. Lee. "Incremental Forming of Mg Alloy Sheet at Elevated Temperatures." *Journal of Mechanical Science and Technology* 21, no. 10 (2007): 1518.
50. Palumbo, G., and M. Brandizzi. "Experimental Investigations on the Single Point Incremental Forming of a Titanium Alloy Component Combining Static Heating with High Tool Rotation Speed." *Materials & Design* 40 (2012): 43–51.
51. Duflou, J.R., Bart Callebaut, Johan Verbert, and Hans De Baerdemaeker. "Laser Assisted Incremental Forming: Formability and Accuracy Improvement." *CIRP Annals* 56, no. 1 (2007): 273–276.
52. Göttmann, Alexander, Jö Diettrich, Georg Bergweiler, Markus Bambach, Gerhard Hirt, Peter Loosen, and Reinhart Poprawe. "Laser-Assisted Asymmetric Incremental Sheet Forming of Titanium Sheet Metal Parts." *Production Engineering* 5, no. 3 (2011): 263–271.
53. Fan, G., L. Gao, G. Hussain, and Z. Wu. "Electric Hot Incremental Forming: A Novel Technique." *International Journal of Machine Tools and Manufacture* 48, no. 15 (2008): 1688–1692.
54. Ambrogio, G., L. Filice, and F. Gagliardi. "Formability of Lightweight Alloys by Hot Incremental Sheet Forming." *Materials & Design* 34 (2012): 501–508.
55. Ji, Y.H., and J.J. Park. "Formability of Magnesium AZ31 Sheet in the Incremental Forming at Hot Temperature." *Journal of Materials Processing Technology* 201, no. 1–3 (2008): 354–358.
56. Ji, Y.H., and J.J. Park. "Incremental Forming of Free Surface with Magnesium Alloy AZ31 Sheet at Hot Temperatures." *Transactions of Nonferrous Metals Society of China* 18 (2008): s165–s169.
57. Ambrogio, G., L. Filice, and G.L. Manco. "Hot Incremental Forming of Magnesium Alloy AZ31." *CIRP Annals* 57, no. 1 (2008): 257–260.
58. Galdos, Lander, Eneko Sáenz de Argandoña, Ibai Ulacia, and Gurutze Arruebarrena. "Hot Incremental Forming of Magnesium Alloys Using Hot Fluid as Heating Media." *Key Engineering Materials* 504 (2012): 815–820.
59. Duflou, J.R., Bart Callebaut, Johan Verbert, and Hans De Baerdemaeker. "Improved SPIF Performance Through Dynamic Local Heating." *International Journal of Machine Tools and Manufacture* 48, no. 5 (2008): 543–549.
60. Callebaut, Bart. "Sheet Metal Forming by Laser Forming and Laser Assisted Incremental Forming." Katholieke Universiteit Leuven: Leuven, The Netherlands, PhD Thesis (2009).
61. Göttmann, A., D. Bailly, Georg Bergweiler, Markus Bambach, Jochen Stollenwerk, Gerhard Hirt, and P. Loosen. "A Novel Approach for Temperature Control in ISF Supported by Laser and Resistance Heating." *The International Journal of Advanced Manufacturing Technology* 67, no. 9–12 (2013): 2195–2205.
62. Mohammadi, Amirahmad, Hans Vanhove, Albert Van Bael, and Joost R. Duflou. "Towards Accuracy Improvement in Single Point Incremental Forming of Shallow Parts Formed Under Laser Assisted Conditions." *International Journal of Material Forming* 9, no. 3 (2016): 339–351.
63. Mohammadi, Amirahmad, Ling Qin, Hans Vanhove, Marc Seefeldt, Albert Van Bael, and Joost R. Duflou. "Single Point Incremental Forming of an Aged AL-Cu-Mg Alloy:

Influence of Pre-Heat Treatment and Hot Forming." *Journal of Materials Engineering and Performance* 25, no. 6 (2016): 2478–2488.
64. Park, J.G., B.S. You, and Y.S. Kim. "A Parametric Study in Incremental Forming of Magnesium Alloy Sheet." *Transactions of Materials Processing* 17, no. 6 (2008): 412–419.
65. Durante, M., A. Formisano, A. Langella, and F. Memola Capece Minutolo. "The Influence of Tool Rotation on an Incremental Forming Process." *Journal of Materials Processing Technology* 209, no. 9 (2009): 4621–4626.
66. Park, Jingee, Jeounghan Kim, Nhokwang Park, and Youngsuk Kim. "Study of Forming Limit for Rotational Incremental Sheet Forming of Magnesium Alloy Sheet." *Metallurgical and Materials Transactions A* 41, no. 1 (2010): 97.
67. Kumar, Ajay, Vishal Gulati, Parveen Kumar, and Hari Singh. "Forming Force in Incremental Sheet Forming: A Comparative Analysis of the State of the Art." *Journal of the Brazilian Society of Mechanical Sciences and Engineering* 41, no. 6 (2019): 251.
68. Xu, Dongkai, Weichao Wu, Rajiv Malhotra, Jun Chen, Bin Lu, and Jian Cao. "Mechanism Investigation for the Influence of Tool Rotation and Laser Surface Texturing (LST) on Formability in Single Point Incremental Forming." *International Journal of Machine Tools and Manufacture* 73 (2013): 37–46.
69. Xu, Dongkai, Bin Lu, Tingting Cao, Jun Chen, Hui Long, and Jian Cao. "A Comparative Study on Process Potentials for Frictional Stir-and Electric Hot-Assisted Incremental Sheet Forming." *Procedia Engineering* 81 (2014): 2324–2329.
70. Davarpanah, Mohammad Ali, Amin Mirkouei, Xiaoyan Yu, Rajiv Malhotra, and Srikanth Pilla. "Effects of Incremental Depth and Tool Rotation on Failure Modes and Microstructural Properties in Single Point Incremental Forming of Polymers." *Journal of Materials Processing Technology* 222 (2015): 287–300.
71. Liu, Zhaobing. "Friction Stir Incremental Forming of AA7075-O Sheets: Investigation on Process Feasibility." *Procedia Engineering* 207 (2017): 783–788.
72. Li, Zhengfang, Shihong Lu, Tao Zhang, Chun Zhang, and Zhixiang Mao. "1060 Al Electric Hot Incremental Sheet Forming Process: Analysis of Dimensional Accuracy and Temperature." *Transactions of the Indian Institute of Metals* 71, no. 4 (2018): 961–970.
73. Fan, Guoqiang, and L. Gao. "Mechanical Property of Ti-6Al-4V Sheet in One-Sided Electric Hot Incremental Forming." *The International Journal of Advanced Manufacturing Technology* 72, no. 5–8 (2014): 989–994.
74. Khazaali, H., and F. Fereshteh-Saniee. "A Comprehensive Experimental Investigation on the Influences of the Process Variables on Hot Incremental Forming of Ti-6Al-4V Titanium Alloy Using a Simple Technique." *The International Journal of Advanced Manufacturing Technology* 87, no. 9–12 (2016): 2911–2923.
75. Liu, Runze, Bin Lu, Dongkai Xu, Jun Chen, Fei Chen, Hengan Ou, and Hui Long. "Development of Novel Tools for Electricity-Assisted Incremental Sheet Forming of Titanium Alloy." *The International Journal of Advanced Manufacturing Technology* 85, no. 5–8 (2016): 1137–1144.
76. Najafabady, S. Amini, and A. Ghaei. "An Experimental Study on Dimensional Accuracy, Surface Quality, and Hardness of Ti-6Al-4 V Titanium Alloy Sheet in Hot Incremental Forming." *The International Journal of Advanced Manufacturing Technology* 87, no. 9–12 (2016): 3579–3588.
77. Van Sy, Le, and Thanh Nam Nguyen. "Hot Incremental Forming of Magnesium and Aluminum Alloy Sheets by Using Direct Heating System." *Proceedings of the Institution of Mechanical Engineers, Part B: Journal of Engineering Manufacture* 227, no. 8 (2013): 1099–1110.
78. Bao, Wenke, Xingrong Chu, Shuxia Lin, and Jun Gao. "Experimental Investigation on Formability and Microstructure of AZ31B Alloy in Electropulse-Assisted Incremental Forming." *Materials & Design* 87 (2015): 632–639.

79. Xu, D.K., B. Lu, T.T. Cao, H. Zhang, J. Chen, H. Long, and Jian Cao. "Enhancement of Process Capabilities in Electrically-Assisted Double Sided Incremental Forming." *Materials & Design* 92 (2016): 268–280.
80. Li, Zhengfang, Shihong Lu, Tao Zhang, Chun Zhang, and Zhixiang Mao. "Electric Assistance Hot Incremental Sheet Forming: An Integral Heating Design." *The International Journal of Advanced Manufacturing Technology* 96, no. 9–12 (2018): 3209–3215.
81. Kinsey, Brad, Graham Cullen, Adam Jordan, and Steven Mates. "Investigation of Electroplastic Effect at High Deformation Rates for 304SS and Ti–6Al–4V." *CIRP Annals* 62, no. 1 (2013): 279–282.
82. Magargee, James, Fabrice Morestin, and Jian Cao. "Characterization of Flow Stress for Commercially Pure Titanium Subjected to Electrically Assisted Deformation." *Journal of Engineering Materials and Technology* 135, no. 4 (2013): 041003.
83. Salandro, Wesley A., Joshua J. Jones, Timothy A. McNeal, John T. Roth, Sung-Tae Hong, and Mark T. Smith. "Formability of Al 5xxx Sheet Metals Using Pulsed Current for Various Heat Treatments." *Journal of Manufacturing Science and Engineering* 132, no. 5 (2010): 051016.
84. Salandro, Wesley A., Joshua J. Jones, Cristina Bunget, Laine Mears, and John T. Roth. *Electrically Assisted Forming: Modeling and Control.* Springer, 2014.
85. Perkins, Timothy A., Thomas J. Kronenberger, and John T. Roth. "Metallic Forging Using Electrical Flow as an Alternative to Hot/Hot Working." *Journal of Manufacturing Science and Engineering* 129, no. 1 (2007): 84–94.
86. Magargee, James, Rong Fan, and Jian Cao. "Analysis and Observations of Current Density Sensitivity and Thermally Activated Mechanical Behavior in Electrically-Assisted Deformation." *Journal of Manufacturing Science and Engineering* 135, no. 6 (2013): 061022.
87. Fan, Guoqiang, Fengtao Sun, Xiangguo Meng, Lin Gao, and Guoquan Tong. "Electric Hot Incremental Forming of Ti-6Al-4V Titanium Sheet." *The International Journal of Advanced Manufacturing Technology* 49, no. 9–12 (2010): 941–947.
88. Zhang, Qinglai, Fugui Xiao, Hailing Guo, Changsheng Li, Lin Gao, Xingwu Guo, Weidong Han, and A.B. Bondarev. "Hot Negative Incremental Forming of Magnesium Alloy AZ31 Sheet: New Lubricating Method." *Journal of Materials Processing Technology* 210, no. 2 (2010): 323–329.
89. Al-Obaidi, Amar, Verena Kräusel, and Dirk Landgrebe. "Hot Single-Point Incremental Forming Assisted by Induction Heating." *The International Journal of Advanced Manufacturing Technology* 82, no. 5–8 (2016): 1163–1171.
90. Al-Obaidi, Amar, Verena Kräusel, and Dirk Landgrebe. "Induction Heating Validation of Dieless Single-Point Incremental Forming of AHSS." *Journal of Manufacturing and Materials Processing* 1, no. 1 (2017): 5.
91. Ambrogio, Giuseppina, Francesco Gagliardi, Ahmad Chamanfar, Wojciech Z. Misiolek, and Luigino Filice. "Induction Heating and Cryogenic Cooling in Single Point Incremental Forming of Ti-6Al-4V: Process Setup and Evolution of Microstructure and Mechanical Properties." *The International Journal of Advanced Manufacturing Technology* 91, no. 1–4 (2017): 803–812.
92. Duflou, Joost R., Anne-Marie Habraken, Jian Cao, Rajiv Malhotra, Markus Bambach, Dave Adams, Hans Vanhove, Amirahmad Mohammadi, and Jack Jeswiet. "Single Point Incremental Forming: State-of-the-Art and Prospects." *International Journal of Material Forming* 11, no. 6 (2018): 743–773.

# 4 Machine Tools and Forming-Tools for Incremental Sheet Forming Process

## 4.1 INTRODUCTION

ISF technology has evolved into an emerging arena of manufacturing during the past number of years. The remarkable progress in the different variants of ISF techniques has inspired researchers and engineers to accurately fabricate various real objects made of various materials. In order to fabricate products, it is customary to develop a platform of machinery and hardware needed to accomplish this die-less process. The set-up of ISF consists of following:

- Forming machinery
- Forming-tools
- ISF set-up
- Lubricants
- CAD/CAM software
- Raw material (to be formed)

## 4.2 FORMING MACHINERY

The forming machinery for accomplishing the ISF process is one of the important elements of hardware needed for this process. It also plays a vital role in the successful forming of components efficiently and accurately. Following parameters of forming machinery determine the performance of the machinery and should be considered in selecting forming machine:

- Stiffness
- Maximum load capacity
- Maximum working area
- Maximum feed rate
- Flexibility and agility
- Cost

The forming machinery for the ISF process can be divided into following three categories:

- CNC milling machines
- Industrial robots
- Purpose-built machines

### 4.2.1 CNC Milling Machines

The CNC milling machine is known for its flexibility, agility and ease of operation for machining processes. It has also been successfully used for the ISF process. The first true ISF process was executed by Mason [1] on a simple lathe machine by clamping the sheet to a spindle. Later, Iseki et al. [2] performed an ISF process on a manually operated three-axis machine, as available CNC milling machines had lower than the required stiffness and accuracy. Also, the CNC milling machine was an expensive machine tool at that time and was not used for forming operations as it was regarded as unsafe. For the ISF process, a CNC milling machine was first used by Iseki et al. [3–5]. Now, CNC milling machines are widely used for the ISF process due to the following:

- High stiffness (approx. 200 kN/mm) [6]
- High load capacity
- Large working area
- Moderate feed-rate
- High flexibility
- High accuracy
- Best suited for SPIF and TPIF processes

Due to several features, described above, CNC milling machines have been attracting researchers and engineers to implement ISF process. The cost of a CNC milling machine is high, which can limit the productivity of the process. However, the high cost of CNC milling can be justified with the fact that apart from the ISF process, other machining operations can be easily performed with a small change in set-up. Moreover, ISF is preferred for prototyping and batch-type production. Hence, the same CNC milling machine can be used for alternative operations (incremental forming as well as machining operations). In addition, CNC milling machines are limited by degrees of freedom (DOFs). CNC milling machines have been used extensively in the literature in order to perform variants of the ISF process. Figure 4.1 shows the CNC milling machine used by Ajay [7]. Nonetheless, the following are some limitations of the CNC milling machine in the ISF process:

- High cost
- Limited DOF

**FIGURE 4.1** Pictorial view of the CNC milling machine set-up for ISF [7].

### 4.2.2 Industrial Robots

The ISF process can also be executed using heavy-duty industrial robots. The use of robots for the ISF process increases the flexibility of the process tremendously, but the lower stiffness of industrial robots limits their forming capacity. However, this problem can be overcome by using heat-assisted ISF (HA-ISF) to form hard-to-form materials (magnesium and titanium alloys) requiring higher forming-forces at room temperature. The principle of using industrial robots for incremental forming has been patented by Tuominen [8]. In this patent, the CNC milling machine has been replaced by an industrial robot with a forming-tool clamped to its wrist. The tool is moved over the surface of the sheet by the robot to form the target geometry. In addition, the patent also proposed a strategy to accomplish the ISF process using two robots. In this strategy, a supporting die is replaced by a moving tool that is controlled by the second robot. At that time, there was no specific name given to this process. This was considered an extension of the TPIF process. Now, this process is known as the DSIF process (as described in Chapter 3). Later, several researchers [9–12] explored the use of industrial robots during the ISF process. Figure 4.2 shows the set-up of synchronized KUKA robots for DSIF.

Using an industrial robot as a forming machine allows the forming-tool to be moved and positioned at various angles relative to the workpiece, enabling the accurate and easy production of complex shapes. A robot has more DOFs than a milling machine, providing better and more accurate control of tool-paths, thus increasing the flexibility of the forming machinery. In addition, a high feed-rate and large working area enable the production of larger components at a faster rate, eliminating the major limitation of the lower production rate of the ISF process.

Now, industrial robots are extensively used for the ISF process due to the following merits:

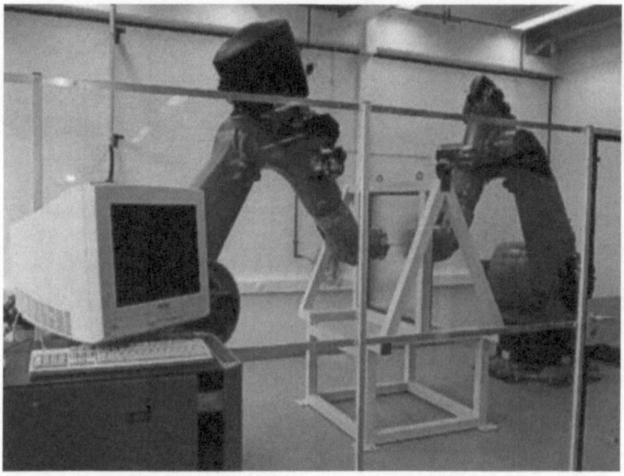

**FIGURE 4.2** A set-up of synchronized KUKA robots for DSIF [13].

- Large working area
- High feed-rate
- High flexibility
- High DOF
- Better control of tool-path
- Suitable for DSIF process

Although industrial robots offer the above advantages for the ISF process, the following drawbacks have been observed by researchers in comparison to CNC milling machines:

- Lower range of stiffness (0.1–120 kN/mm) [6]
- Less load capacity [13]
- Less accuracy [14]

The lower stiffness and lower load capacity of the robots result in the inaccuracy during the positioning of the forming-tool and workpiece, which directly leads to a loss of dimensional integrity of the components [14]. Some researchers [15, 16] have found that the industrial robots need less forming-force to deform the material as compared to that needed by CNC milling machines. This allows industrial robots to function with less power and makes the ISF process economical.

### 4.2.3 Purpose-Built Machine

CNC milling machines and industrial robots are general-purpose machine tools for performing multiple tasks and ISF can also be done using these machines. ISF, being a special forming technique, needs a forming-tool that can form the material layer by layer locally [17, 18]. The following are some key requirements of ISF that demand modifications in the forming machinery [19]:

- Forming-forces, dimensions, strains and formability need to be measured dynamically during the ISF process.
- Requirement of different DOFs of the machine depending on the shape of the product.
- Access to the back side of workpiece.
- Change of forming-tool or forming tooltips with ease.
- Need for high load capacity normal to the spindle direction.

In some variants of ISF (like DSIF and TPIF) there is a need to change the DOF of the machine and to access the reverse side of the workpiece for the forming operation. In addition, the load capacities of the CNC milling machines and industrial robots are lower, which may result in damage to the machinery as high load capacity is required during forming operations [14, 19]. Due to limited DOFs and load-carrying capacity, CNC milling machines and industrial robots limit the applications of the ISF process in terms of design and materials. In ISF, the material's deformation is different from that in cutting operations. For example, springback is experienced during ISF operations, leading to dimensional inaccuracy [20].

In order to meet the aforementioned requirements for improving ISF, the following should be implemented to adapted CNC milling machines or industrial robots:

- The worktable must be redesigned to clamp ISF fixture [20].
- Tool-path correction algorithms must be incorporated in order to avoid large geometrical deviations [9].
- Tool-holders must be reinforced to avoid lateral deformation in the spindle.
- Cumulative and positioning errors must be eliminated.
- The forming machine should be equipped with measurement instruments that can measure response and provide feedback to correct tool-paths.
- Load capacity must be increased in the axial direction of the forming-tool for safe forming of hard-to-form materials.
- The mounting of the forming-tool and change of the forming-tool or tool-tip should be quick and easy.

Considering the above factors, special machines for ISF, called purpose-built machines, have been developed:

a) Amino ISF machine
b) Cambridge ISF machine
c) SPIF-A machine

(a) *Amino ISF machine*: In 1996, AMINO Corporation developed a dedicated machine tool for ISF. This specially designed machine tool, shown in Figure 4.3, is equipped with a pneumatic-controlled work table that is capable of moving for pre-stretching the work material as required in hybrid forming. In 1998, AMINO Corporation also proposed CAM software for this purpose-built machine. In 2002, AMINO Corporation started

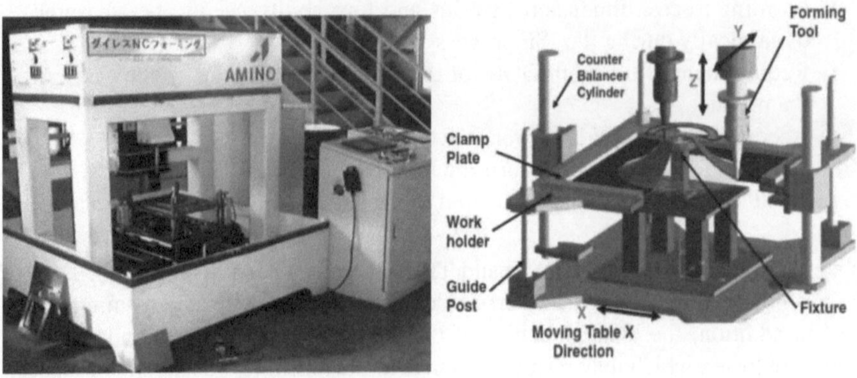


(a) (b)

**FIGURE 4.3** Amino ISF machine: (a) first prototype developed in 1996, (b) schematic of first prototype [22].

to supply these purpose-built machines commercially [21, 22]. The Amino ISF machine has the following features:

- Better accuracy than adapted CNC milling machines and industrial robots
- Higher load capacity in the axial direction of the forming-tool
- Equipped with partial die set-up as required in TPIF

This purpose-built machine with the above features is suitable for the TPIF process with ease and accurately. However, the SPIF operation cannot be performed on it as the partial die is fixed on its table, although there are possibilities to change the set-up and to equip this machine with online force, temperature and strain measurement devices [20].

(b) *Cambridge ISF machine*: After AMINO, the University of Cambridge developed a purpose-built machine for the SPIF process in 2004. This special ISF machine, known as the "*Cambridge ISF machine*", was designed by Allwood et al. [19]. This machine had the following features that distinguish it from the Amino ISF machine:

- Online measurement and estimation of tool forces
- Simple change of tool-tips
- Access to the reverse side of the workpiece

Moreover, the machine was also designed to predict the forming-force for different materials. Later, the following features were added to the Cambridge ISF machine:

- Online feedback to control the tool-path using algorithms
- Design of slave-tool (which acts from the back side of the workpiece)

This machine has a payload of 13 kN and 6.5 kN in the axial and lateral directions of the forming-tool, respectively. The working area of this machine is 500 × 500 mm$^2$ and the maximum travel of the vertical axis is 400 mm. The Cambridge ISF machine is suitable for performing DSIF and SPIF using the same set-up. It also increases the accuracy of the parts as it is equipped with an online tool-path correction system using feedback and algorithms. It also has constraints in positioning flexibility of the forming-tool due to the involvement of a large tapered spindle [20].

(c) *SPIF-A machine*: This purpose-built machine was designed and fabricated at the University of Aveiro for performing the SPIF operation. The schematic of this purpose-built machine is shown in Figure 4.4. A special kinematic system (also known as the *Stewart platform*) having six DOFs provides positioning flexibility to the forming-tool. This kinematic system allows the machine to have a payload of 13 kN and 6.5 kN in the axial and lateral directions of the forming-tool, respectively. The working area of this machine is 300 × 300 mm$^2$ and the maximum travel of the vertical axis is 100 mm. The following are some key features of this purpose-built machine [20]:

- Six DOFs for the forming-tool allow improved accuracy and flexibility.
- The Stewart platform provides extra stiffness to the forming-tool and machine components.
- Load-cells provided between spindle and kinematic system allow the online measurement of the forming-force and in-plane bending moment.
- The Stewart platform minimizes positioning errors.

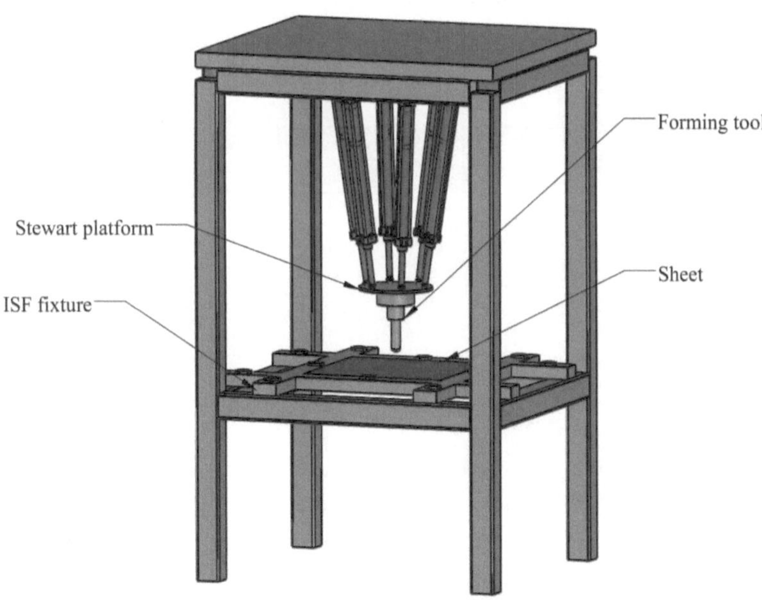


**FIGURE 4.4** Schematic of SPIF-A machine.

## 4.3 ISF FORMING-TOOLS

The forming-tool geometry is an important process parameter in the ISF process that affects the forming-force, surface roughness, strains and formability of the formed parts. In ISF, a simple and generic forming-tool can produce infinite varieties of shapes in components [23, 24]. In ISF, the geometry of the forming-tool has a direct impact on the forming-forces required to deform the material. The following are the common types of forming-tools used in ISF:

- Hemispherical end
- Flat end
- Vertical roller-ball end
- Oblique roller-ball end

For the ISF process, the forming-tool is normally a hemispherical-end (Figure 4.5a) or flat-end (Figure 4.5b) steel bar. These shapes generate more heat due to friction and sometimes produce scratches on the finished part. In order to eliminate this problem, Durante et al. [25] introduced a vertical roller-ball (VRB) tool (Figure 4.5) that reduced the friction and enhanced the quality of the finished part. This VRB tool has limitation to produce higher wall angle of the formed components due to the interference between the VRB tool shank and workpiece (see Figure 4.5c). Hence, Lu et al. [26] proposed an oblique roller-ball (ORB)-tipped tool as shown in Figure 4.5d, which removed the limitation of the maximum wall angle. There is limited reported research for the use of the roller-ball tool for the ISF process. The findings of researchers who have investigated the effects of different types of forming-tools on forming-forces are summarized in Table 4.1.

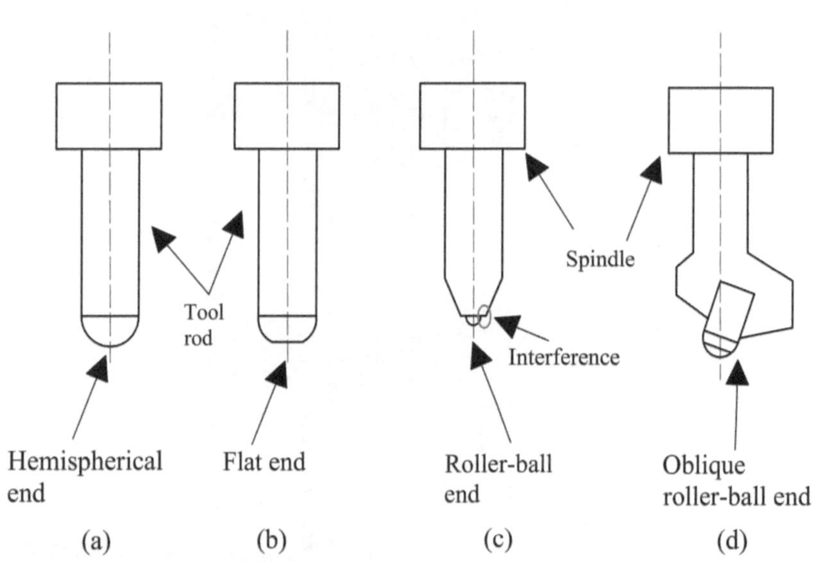


**FIGURE 4.5** Geometry of ISF forming-tool ends.

**TABLE 4.1**
**Summary of Effects of Different Types of ISF Forming-Tools**

| Sr. No. | Types and Parameters Studied | Tool Material | Sheet Material | Findings | Reference |
|---|---|---|---|---|---|
| 1 | Flat-end tools and hemispherical-end tools of different diameters and wall angles | High-speed steel | AA6063 | When the tool shape was changed from flat-end to hemispherical-end, the maximum axial forces decreased approximately 14.17, 17.46 and 14.69% for 7.52, 11.60 and 15.66 mm tool diameters, respectively. | [17] |
| 2 | Hemispherical-end tools (diameter 10, 20, 30 mm) and rollerball tool (diameter 25.4 mm) | Tungsten carbide tip and K110 steel body | AA7075-O | Forming-force increased after each loop of the tool. The maximum force was 2700 N in the vertical direction and 1800 N in the horizontal direction. | [26] |
| 3 | Flat-end tools and hemispherical-end tools of different diameters | High-speed steel | AA6063, AA2024-O | The lower axial peak forces were required to deform the sheet material when a hemispherical-end tool was employed. The flat-end tools produced larger forming-forces. | [27] |
| 4 | Flat-end tools of diameter 10 mm and hemispherical-end tools of diameter 4, 6 and 10 mm | High-speed steel | AA 3003-O | Hemispherical-end tool required greater forming-force than the flat-end tool. These results are contradictory to [27]. | [28] |
| 5 | Hemispherical-end, VRB and ORB tools of diameter 10 mm | - | AA1100, AA2024, AA5052, AA6111 | Surface roughness was found to be in decreasing order when forming-tool was changed from hemispherical-end to VRB through ORB. Horizontal forces by the hemispherical-end tool were greater than those required by the ORB tool. The difference between vertical forces produced by the hemispherical-end and ORB tools was negligible. | [29] |

**FIGURE 4.6** Set-up for ISF [27].

Kumar et al. [30] found that the formability of the components increased when a hemispherical-end tool was employed in place of a flat-end tool. Some researchers [31, 32] have also investigated the effects of tool shape on the surface roughness of components and found that the surface roughness of the components formed by a hemispherical-end tool was less than those formed by a flat-end tool.

## 4.4 TYPICAL ISF SET-UP

The set-up for the ISF process consists of a specially designed hollow fixture to clamp the blank rigidly and constrain the motion of the blank in all DOFs. The hollow fixture, also known as the ISF fixture, is mounted on the table of the machine tool (CNC milling machine/robotic set-up/purpose-built machine). The forming-tool is firmly clamped into the spindle of the forming machinery. The ISF set-up is shown in Figure 4.6. A force dynamometer can be clamped between the ISF fixture and the table of the forming machinery to record online the forming-force. After the completion of the forming process, the formed component is removed from the fixture.

The ISF fixture is essential in the successful and safe execution of the ISF operation. This makes the design and fabrication of the ISF fixture crucial. The size of the fixture depends on the size of the part to be formed and on the capacity of the forming machinery. The ISF fixture consists of the following components:

- Base plate or static frame
- Static columns

- Top plate
- Backing plate
- Clamping plate

The exploded view of the arrangement of these components is shown in Figure 4.7.

*Base plate*: It is the static plate that is fixed to the machine table, and the top plate is fixed to it with static columns. The space between the top plate and the base plate is the hollow space below the blank, for forming.

*Static columns*: These are the supports for the top plate. The height of the static columns depends upon the depth of the part to be formed.
Top plate: This plate is fixed to the base plate with static columns and is between the base plate and clamping plate. The blank and backing plate are placed on it.

*Backing plate*: The backing plate acts as a die in order to obtain the required shape in the ISF process. For example, to produce a conical frustum, the backing plate will have a circular hollow section of the required size. Likewise, it will have a square

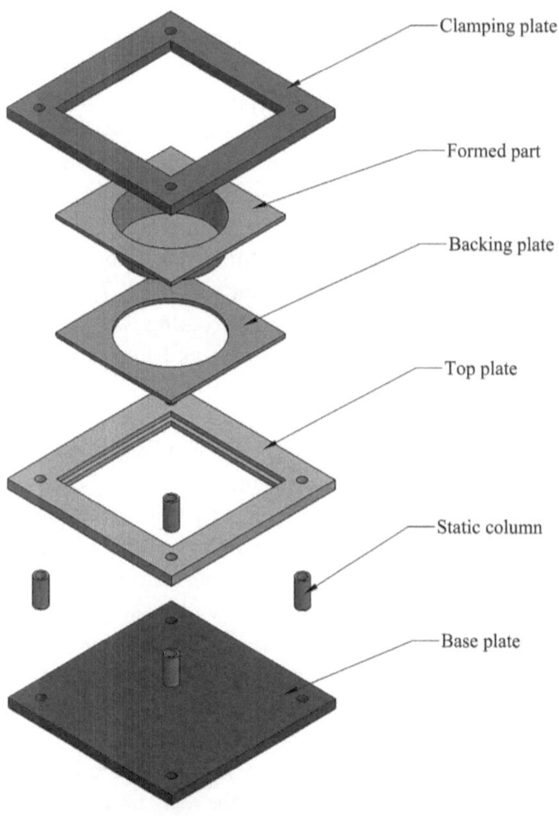


**FIGURE 4.7** Exploded view of ISF fixture.

hollow section to produce a pyramidal shape. The sheet material (blank) is placed on the backing plate and clamped by the clamping plate. The backing plate of the ISF fixture avoids unwanted bending of the sheet during the initial phase of the forming operation and allows the forming of the part with increased geometrical accuracy.

*Clamping plate*: It is used to clamp the sheet blank into the fixture using a properly designed clamping system. The ISF fixture design is complex as it should properly clamp the sheet blank and not allow unwanted deformations with the quick clamping and removal of the formed part.

## 4.5 LUBRICANTS

The material to material friction between the tool-tip and the workpiece surface greatly affects product quality, tool wear, deformation mechanics and required forming-forces, which in turn impact the production cost. In the ISF process, lubrication is used to reduce the friction between the tool-tip and the workpiece and can be applied in the form of the paste or spray coating of the blank to produce self-lubricating effects or in liquid form with the forming-tool and blank submerged in lubricant. In order to ensure the presence of lubricant between the tool and blank during the ISF process, the lubricant must be able to withstand forming temperatures and should not get squeezed out [33]. Lubrication also affects the surface quality, defects and formability of the formed component. The right choice of lubricant can reduce friction at the tool–sheet interface leading to less forming-forces during the ISF process, increasing heat dissipation and lowering tool wear, which in turn increase the tool life [34–36]. A reduction in the forming-forces requires smaller and economical machines and the optimization of the material flow by reducing the number of stages necessary for forming the final shape. Moreover, homogenous deformation and reduced internal residual stresses informed parts may be achieved by efficient lubrication. Most of the lubricants behave as thermal insulators at the contact zone which increases the tool life [37].

Hussain et al. [38] claimed that the method of coating is not suitable for larger values (greater than 40) of the ratio of tool diameter and step down. Some researchers [39–41] have claimed that mineral oils with low viscosity cannot withstand forming temperatures and squeeze out.

Sometimes, it becomes difficult to form a defect-free surface of a component due to adhesion between the forming-tool and workpiece because binding forces between the molecules of the surfaces in contact predominate the physical properties of the material. Lubricants should also be able to restrict adhesion and abrasion at the tool–sheet interface. The viscosity of the lubricant also plays a crucial role at the contact zone between the forming-tool and workpiece [42]. Tribological effects are predominant in sheet metal processes [43]. For the high forming-force at the tool–sheet interface in ISF applications, a highly viscous lubricant is useful. Highly viscous oil does not squeeze out easily from the contact zone which helps in maintaining separation between the forming-tool and sheet [37].

Several researchers have investigated the use of different types of lubricants for different workpiece shapes produced by the ISF process, as summarized in Table 4.2. Excellent thermal stability, oxidation resistance, good water resistance, de-emulsification characteristics, rust resistance and copper corrosion protective

**TABLE 4.2**
**Summary of Lubricants, Investigated in the Literature, for the ISF Process**

| Sr. No. | Lubricant | Sheet Material | Work-Piece Shape | Reference |
|---|---|---|---|---|
| 1 | Repsol SAE30, Finarol B5746, Moly-Slip AS40, Welcone AL-M, Moly-Slip HSB | DP780, AA1050-T4 | Conical frustum | [36] |
| 2 | Shell Tellus oil 68 | AA7075-O | Conical frustum | [44–46] |
| 3 | Grease | AA5086 H-111, AA1050A, DC04, TiCPGR.2, FeP04 steel | Conical frustum, pyramidal frustum, dome shape | [15, 16, 47, 48] |
| 4 | Castrol oil Alpha SP 68, Alpha SP 150, Alpha SP 320 | AA2024-O, AA6063-O | Conical frustum | [7, 17, 18, 27, 30–32, 40, 41] |
| 5 | Blasocut BC 20, Nuto H46, Febis K220, grease | AA3003-O | Conical frustum, pyramidal frustum | [49] |
| 6 | Houghton TD-52 | AISI304 steel | Conical frustum | [50, 51] |
| 7 | Sylac 80-05 | DC05 steel | Conical frustum | [52] |
| 8 | Coating of Vaseline | AA5052 | Conical frustum, pyramidal frustum | [53] |
| 9 | Rocol RTD compound | TA1 titanium | Skull-shaped cranial plate | [54] |
| 10 | PTFE-based grease | PVC, PLA | Conical frustum | [55] |
| 11 | Graphite particles | AA5052, DDQ steel | Conical frustum | [56] |
| 12 | Industrial oil | AA1050A | Conical frustum | [57] |
| 13 | Machine oil | AA3003-O AA1100, AA2024, AA5052 | Flower shape, conical frustum | [28, 29] |
| 14 | Forming oil | AA7075-T0, AA6016, AA3003-O, DC04 steel | Pyramid frustum, grove shape, conical frustum, dome | [25, 58–60] |

nature of lubricants result in less maintenance and a longer life of the lubricant and forming-tool [7, 14, 27, 31].

## 4.6 CAD/CAM SOFTWARE

Once the set-up is ready for the execution of the ISF process, the next step is to provide the required NC instructions to the forming machinery (CNC milling machine/industrial robot/purpose-built machine). First of all, a 2D/3D model of the desired shape of the component is created using any suitable CAD software (like Solidworks®/Catia®/Pro-E®/AutoCAD®, etc.). Later, the CAD model is exported to suitable CAM software to generate the NC instructions. These NC instructions control the path of the forming-tool, step size, feed rate and other functions of the forming machinery (like tool-rotation, heating and lubrication). Figure 4.8 is the procedure for developing a CAD/CAM environment for the ISF process. The path traversed by the ISF forming-tool, relative to the workpiece surface, is chosen in the CAM software. The most commonly used paths are profile and helical tool-paths. Commercial CAM packages for milling machines offer different tool-paths for the forming and machining applications. A suitable tool-path is chosen in order to fulfill the requirement of the target geometry of the product. A detailed explanation of different types of tool-paths used in the ISF process will be given in Chapter 5. The NC instructions are generally created in the form of G-codes and M-codes using CAM packages. These NC instruction files are loaded in the forming machinery to accomplish the ISF process precisely. Various researchers have formed different types of shapes using this approach as shown in Table 4.3.

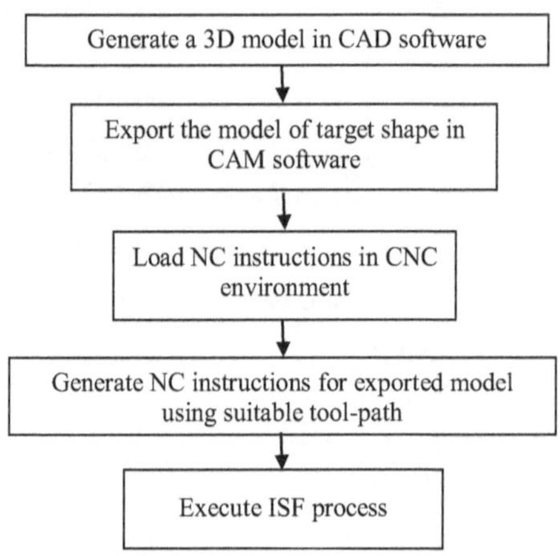


**FIGURE 4.8** Procedure for developing CAD/CAM environment for the ISF process.

## 4.7 RAW MATERIALS FOR ISF

Aluminum and steel alloy parts are commonly formed using the ISF process. Some researchers have also investigated the forming of polyvinyl chloride (PVC) [69],

**TABLE 4.3**
**Summary of the Geometries and Shapes of Various Parts Investigated in the Literature of the ISF Process**

| Target Shape | Material | ISF Machinery Used | Reference |
|---|---|---|---|
| Conical frustum | DP780, AA1050-T4, AA7075-O | Purpose-built machine | [36, 44, 45] |
| Conical frustum | AA5086-H11, AA2024-T3, AA2024-O, AA2024-W, AA5182, | Industrial robots | [15, 16, 61, 62] |
| Pyramidal frustum | St14 steel | | [63] |
| Pyramidal frustum | DC04, DP1000, 22MnB5, Ti-6Al-4V, AA5052-T0, FPG copper, AA1060, AA3003-O, PVC, AA7075-T0 | CNC milling machine | [25, 49, 64–70] |
| Conical frustum | Magnesium AZ31B, Ti-6Al-4V, Castrol SP320, AA2024-O, AA6063-O, DC04 steel, AA3003-O, AA7075-O, AA1050-O, DC05 steel, PVC, PLA | CNC milling machine | [7, 17, 18, 27, 30–32, 40, 41, 46, 52, 55, 59, 71–74] |
| Dome | DC04 steel, AA 1050A, DC04, FeP04 steel | CNC milling machine | [47, 48, 60] |
| Groove | AA6016, AA7075-O | CNC milling machine | [26, 58] |
| Flower | AA1100, AA2024, AA5052, AA6111 | CNC milling machine | [29] |
| Skull-shaped cranial plate | TA1 titanium | CNC milling machine | [54] |

magnesium [55, 64, 71] and copper alloy [66] sheets. Since ISF has many possible medical applications, biocompatible materials like stainless steel, cobalt and chromium alloys and titanium and titanium alloys [54, 65, 72] are also investigated. Aluminum alloys are common in the aero and space sector. The forming of aluminum alloys is executed from very low forming temperature to elevated temperature. Tables 4.1, 4.2 and 4.3 show different materials investigated by researchers for the ISF process.

It is important to know the chemical compositions and physical properties of the material of the sheet to be formed using ISF to determine the feasibility of safe forming and to propose some guidelines for the failure mechanism of the material relative to the formability, forming-force and stress–strain patterns.

## 4.8 CONCLUDING REMARKS

The precision and quality of machine tools used for any manufacturing process determine the quality and cost of the produced components. The ISF process requires the use of advanced machinery and special forming-tools, and this die-less process requires a CAD/CAM environment. Hence, various software and tool-path optimization techniques become crucial for enhancing the effectiveness and reliability of the ISF process. According to the requirements of products and their material, various researchers have employed CNC milling machines, industrial robots and purpose-built machines and have considered several factors like stiffness, rigidity, load capacity, feed rate and the agility of the forming machinery. Purpose-built machines enhance the agility and flexibility of the process whereas industrial robots speedup the forming operation for increasing the productivity of the process. Based on the published literature, it can be inferred that the forming-tool plays a vital role in the successful forming of components by the ISF process, and, hence, it should be selected carefully. The ISF set-up should be rigid enough in order to constrain the motion between the sheet and ISF fixture. The use of a backing plate in ISF fixture can enhance the geometrical accuracy of the components at the expense of the agility of the process. The lubrication aspects of the ISF process cannot be ignored. The right choice of lubricant results in less maintenance of machinery, smaller forming-forces and longer life of the forming-tool. Researchers have urged that thermal stability, oxidation resistance, de-emulsification characteristics, rust resistance and copper corrosion protective nature of lubricants should be taken into account.

## REFERENCES

1. Mason, B. "Sheet Metal Forming for Small Batches." *Bachelor Thesis*, University of Nottingham, Nottingham, UK, 1978.
2. Iseki, H., K. Kato, and S. Sakamoto. "Flexible and Incremental Sheet Metal Forming Using a Spherical Roller." *Proceedings of 40th JJCTP* 41 (1989): 44.
3. Iseki, H. "Flexible and Incremental Sheet Metal Bulging Using a Path-Controlled Spherical Roller (A Trial Manufacture of the Bulging Machine)." *Proceedings of JSCTP* (1991): 515–518 (in Japanese).
4. Iseki, H. "Flexible and Incremental Sheet Metal Bulging Using a Path Controlled Spherical Roller." *Transactions of the Japan Society of Mechanical Engineers C* 58, no. 554 (1992a): 3147.
5. Iseki, H. "Flexible and Incremental Sheet Metal Bulging Using a Few Spherical Rollers." *Transactions of the Japan Society of Mechanical Engineers* 59 (1993): 2849.
6. Behera, Amar Kumar, Ricardo Alves de Sousa, Giuseppe Ingarao, and Valentin Oleksik. "Single Point Incremental Forming: An Assessment of the Progress and Technology Trends from 2005 to 2015." *Journal of Manufacturing Processes* 27 (2017): 37–62.
7. Ajay. "Optimization of Process Parameters in Incremental Sheet Forming: Experimental and Theoretical Analysis." *PhD Thesis* (2018).
8. Tuominen, Timo. "Method and Apparatus for Forming Three-Dimensional Shapes in a Sheet Metal." Patent number WO2004030843A1, 2004.
9. Meier, Horst, V. Smukala, O. Dewald, and Jian Zhang. "Two Point Incremental Forming with Two Moving Forming Tools." *Key Engineering Materials* 344 (2007): 599–605.
10. Wang, Yongjun, Ying Huang, Jian Cao, and N. Venkata Reddy. "Experimental Study on a New Method of Double Side Incremental Forming." In: *ASME 2008 International*

*Manufacturing Science and Engineering Conference Collocated with the 3rd JSME/ASME International Conference on Materials and Processing*, October 7–10, 2008, Evanston, Illinois, USA, pp. 601–607. American Society of Mechanical Engineers Digital Collection, 2009.
11. Franzen, V., L. Kwiatkowski, G. Sebastiani, R. Shankar, A.E. Tekkaya, and M. Kleiner. "Dyna-Die: Towards Full Kinematic Incremental Forming." *International Journal of Material Forming* 1, no. 1 (2008): 1163–1166.
12. Kwiatkowski, Lukas, Miroslav Urban, Gerd Sebastiani, and A. ErmanTekkaya. "Tooling Concepts to Speed Up Incremental Sheet Forming." *Production Engineering* 4, no. 1 (2010): 57–64.
13. Meier, H., B. Buff, R. Laurischkat, and V. Smukala. "Increasing the Part Accuracy in Dieless Robot-Based Incremental Sheet Metal Forming." *CIRP Annals* 58, no. 1 (2009): 233–238.
14. Kumar, Ajay, Vishal Gulati, Parveen Kumar, and Hari Singh. "Forming Force in Incremental Sheet Forming: A Comparative Analysis of the State of the Art." *Journal of the Brazilian Society of Mechanical Sciences and Engineering* 41, no. 6 (2019): 251.
15. Belchior, Jérémy, Dominique Guines, Lionel Leotoing, and Eric Ragneau. "Force Prediction for Correction of Robot Tool Path in Single Point Incremental Forming." *Key Engineering Materials* 554 (2013): 1282–1289.
16. Belchior, Jérémy, Lionel Leotoing, Dominique Guines, Eric Courteille, and Patrick Maurine. "A Process/Machine Coupling Approach: Application to Robotized Incremental Sheet Forming." *Journal of Materials Processing Technology* 214, no. 8 (2014): 1605–1616.
17. Kumar, Ajay, Vishal Gulati, and Parveen Kumar. "Experimental Investigation of Forming Forces in Single Point Incremental Forming." In: *Advances in Industrial and Production Engineering*, Shanker, Kripa, Shankar, Ravi, Sindhwani, Rahul (Eds.), pp. 423–430. Springer, Singapore, 2019.
18. Gulati, Vishal, and A. Kumar. "Investigation of Some Process Parameters on Forming Force in Single Point Incremental Forming." *International Research Journal of Engineering and Technology* 4, no. 4 (2018): 784–791.
19. Allwood, J.M., N.E. Houghton, and K.P. Jackson. "The Design of an Incremental Sheet Forming Machine." *Advanced Materials Research* 6 (2005): 471–478.
20. de Sousa, R.J. Alves, J.A.F. Ferreira, J.B. Sá de Farias, J.N.D. Torrão, D. Afonso, and M.A.B.E. Martins. "SPIF-A: On the Development of a New Concept of Incremental Forming Machine." *Structural Engineering and Mechanics* 59, no. 5 (2014): 645–660.
21. Amino, Hiroyuki, Yan Lu, Trent Maki, S. Osawa, and K. Fukuda. "Dieless NC Forming, Prototype of Automotive Service Parts." In: *Proceedings of the 2nd International Conference on Rapid Prototyping and Manufacturing (ICRPM)*, Beijing, 2002.
22. Amino, Masaaki, Masashi Mizoguchi, Yuji Terauchi, and Trent Maki. "Current Status of "Dieless" Amino's Incremental Forming." *Procedia Engineering* 81 (2014): 54–62.
23. Li, Yanle, Xiaoxiao Chen, Zhaobing Liu, Jie Sun, Fangyi Li, JianfengLi, and Guoqun Zhao. "A Review on the Recent Development of Incremental Sheet-Forming Process." *The International Journal of Advanced Manufacturing Technology* 92, no. 5–8 (2017): 2439–2462.
24. Dittrich, Marc-André, T.G. Gutowski, Jian Cao, John Timothy Roth, Z.C. Xia, V. Kiridena, F. Ren, and H. Henning. "Exergy Analysis of Incremental Sheet Forming." *Production Engineering* 6, no. 2 (2012): 169–177.
25. Durante, M., A. Formisano, and A. Langella. "Observations on the Influence of Tool-Sheet Contact Conditions on an Incremental Forming Process." *Journal of Materials Engineering and Performance* 20, no. 6 (2011): 941–946.
26. Li, Yanle, Zhaobing Liu, W.J.T. Daniel, and P.A. Meehan. "Simulation and Experimental Observations of Effect of Different Contact Interfaces on the

Incremental Sheet Forming Process." *Materials and Manufacturing Processes* 29, no. 2 (2014): 121–128.
27. Kumar, Ajay, and Vishal Gulati. "Experimental Investigations and Optimization of Forming Force in Incremental Sheet Forming." *Sādhanā* 43, no. 10 (2018): 159.
28. Ziran, Xu, L. Gao, G. Hussain, and Z. Cui. "The Performance of Flat End and Hemispherical End Tools in Single-Point Incremental Forming." *The International Journal of Advanced Manufacturing Technology* 46, no. 9–12 (2010): 1113–1118.
29. Lu, B., Y. Fang, D.K. Xu, J. Chen, H.Ou, N.H. Moser, and Jian Cao. "Mechanism Investigation of Friction-Related Effects in Single Point Incremental Forming Using a Developed Oblique Roller-Ball Tool." *International Journal of Machine Tools and Manufacture* 85 (2014): 14–29.
30. Kumar, Ajay, Vishal Gulati, Parveen Kumar, Vinay Singh, Brijesh Kumar, and Hari Singh. "Parametric Effects on Formability of AA2024-O Aluminum Alloy Sheets in Single Point Incremental Forming." *Journal of Materials Research and Technology* 8, no. 1 (2019): 1461–1469.
31. Kumar, Ajay, and Vishal Gulati. "Experimental Investigation and Optimization of Surface Roughness in Negative Incremental Forming." *Measurement* 131 (2019): 419–430.
32. Kumar, Ajay, Vishal Gulati, and Parveen Kumar. "Effects of Process Parameters on Surface Roughness in Incremental Sheet Forming." *Materials Today: Proceedings* 5, no. 14 (2018): 28026–28032.
33. Zhang, Qinglai, Fugui Xiao, Hailing Guo, Changsheng Li, Lin Gao, Xingwu Guo, Weidong Han, and A.B. Bondarev. "Warm Negative Incremental Forming of Magnesium Alloy AZ31 Sheet: New Lubricating Method." *Journal of Materials Processing Technology* 210, no. 2 (2010): 323–329.
34. Tisza, Miklós, and Tibor Fülöp. "A General Overview of Tribology of Sheet Metal Forming." *Journal for Technology of Plasticity* 26, no. 2 (2001): 11–26.
35. Lee, B.H., Y.T. Keum, and R.H. Wagoner. "Modeling of the Friction Caused by Lubrication and Surface Roughness in Sheet Metal Forming." *Journal of Materials Processing Technology* 130 (2002): 60–63.
36. Azevedo, Nelson Gil, João Sá Farias, Ricardo Pereira Bastos, Pedro Teixeira, João Paulo Davim, and Ricardo Jose Alves de Sousa. "Lubrication Aspects During Single Point Incremental Forming for Steel and Aluminum Materials." *International Journal of Precision Engineering and Manufacturing* 16, no. 3 (2015): 589–595.
37. Wilson, W.R.D. "Friction and Lubrication in Bulk Metal-Forming Processes." *Journal of Applied Metalworking* 1, no. 1 (1978): 7–19.
38. Hussain, Gao, L. Gao, N. Hayat, Z. Cui, Y.C. Pang, and N.U. Dar. "Tool and Lubrication for Negative Incremental Forming of a Commercially Pure Titanium Sheet." *Journal of Materials Processing Technology* 203, no. 1–3 (2008): 193–201.
39. Kim, Y.H., and J.J. Park. "Effect of Process Parameters on Formability in Incremental Forming of Sheet Metal." *Journal of Materials Processing Technology* 130 (2002): 42–46.
40. Kumar, Ajay, Vishal Gulati, and Parveen Kumar. "Investigation of Surface Roughness in Incremental Sheet Forming." *Procedia Computer Science* 133 (2018): 1014–1020.
41. Kumar, Ajay, Vishal Gulati, and Parveen Kumar. "Investigation of Process Variables on Forming Forces in Incremental Sheet Forming." *International Journal of Engineering and Technology* 10 (2018): 680–684.
42. Singer, Markus, and Mathias Liewald. "Effect of Surface Enlargement and of Viscosity of Lubricants on Friction Behaviour of Advanced High Strength Steel Material During Deep Drawing." *Advanced Materials Research* 1018 (2014): 253–260.
43. Blake, T.D., and Y.D. Shikhmurzaev. "Dynamic Wetting by Liquids of Different Viscosity." *Journal of Colloid and Interface Science* 253, no. 1 (2002): 196–202.

44. Li, Yanle, William JT Daniel, Zhaobing Liu, Haibo Lu, and Paul A. Meehan. "Deformation Mechanics and Efficient Force Prediction in Single Point Incremental Forming." *Journal of Materials Processing Technology* 221 (2015): 100–111.
45. Li, Yanle, Zhaobing Liu, Haibo Lu, W.J.T. Bill Daniel, Sheng Liu, and Paul A. Meehan. "Efficient Force Prediction for Incremental Sheet Forming and Experimental Validation." *The International Journal of Advanced Manufacturing Technology* 73, no. 1–4 (2014): 571–587.
46. Liu, Zhaobing, Yanle Li, and Paul Anthony Meehan. "Experimental Investigation of Mechanical Properties, Formability and Force Measurement for AA7075-O Aluminum Alloy Sheets Formed by Incremental Forming." *International Journal of Precision Engineering and Manufacturing* 14, no. 11 (2013): 1891–1899.
47. Fiorentino, A. "Force-Based Failure Criterion in Incremental Sheet Forming." *The International Journal of Advanced Manufacturing Technology* 68, no. 1–4 (2013): 557–563.
48. Fiorentino, Antonio, Elisabetta Ceretti, Aldo Attanasio, Luca Mazzoni, and Claudio Giardini. "Analysis of Forces, Accuracy and Formability in Positive Die Sheet Incremental Forming." *International Journal of Material Forming* 2, no. 1 (2009): 805.
49. Duflou, Joost, Yasemin Tunckol, Alex Szekeres, and Paul Vanherck. "Experimental Study on Force Measurements for Single Point Incremental Forming." *Journal of Materials Processing Technology* 189, no. 1–3 (2007): 65–72.
50. Centeno, Gabriel, Isabel Bagudanch, Andrés Jesús Martínez-Donaire, Maria Luisa Garcia-Romeu, and Carpóforo Vallellano. "Critical Analysis of Necking and Fracture Limit Strains and Forming Forces in Single-Point Incremental Forming." *Materials & Design* 63 (2014): 20–29.
51. Bagudanch, I., G. Centeno, C. Vallellano, and M.L. Garcia-Romeu. "Forming Force in Single Point Incremental Forming Under Different Bending Conditions." *Procedia Engineering* 63 (2013): 354–360.
52. Petek, Aleš, Karl Kuzman, and Janez Kopač. "Deformations and Forces Analysis of Single Point Incremental Sheet Metal Forming." *Archives of Materials science and Engineering* 35, no. 2 (2009): 107–116.
53. Wang, Jyhwen, Mahesh Nair, and Ying Zhang. "An Efficient Force Prediction Strategy in Single Point Incremental Sheet Forming." *Procedia Manufacturing* 5 (2016): 761–771.
54. Lu, B., H. Ou, S.Q. Shi, H. Long, and J. Chen. "Titanium Based Cranial Reconstruction Using Incremental Sheet Forming." *International Journal of Material Forming* 9, no. 3 (2016): 361–370.
55. Davarpanah, Mohammad Ali, Amin Mirkouei, Xiaoyan Yu, Rajiv Malhotra, and Srikanth Pilla. "Effects of Incremental Depth and Tool Rotation on Failure Modes and Microstructural Properties in Single Point Incremental Forming of Polymers." *Journal of Materials Processing Technology* 222 (2015): 287–300.
56. Malhotra, Rajiv, Jian Cao, Michael Beltran, Dongkai Xu, James Magargee, Vijitha Kiridena, and Z. Cedric Xia. "Accumulative-DSIF Strategy for Enhancing Process Capabilities in Incremental Forming." *CIRP Annals* 61, no. 1 (2012): 251–254.
57. Shrivastava, Parnika, and Puneet Tandon. "Investigation of the Effect of Grain Size on Forming Forces in Single Point Incremental Sheet Forming." *Procedia Manufacturing* 2 (2015): 41–45.
58. Blaga, Adrian, Octavian Bologa, Valentin Oleksik, and Bogdan Pirvu. "Experimental Researches Regarding the Influence of Geometric Parameters on the Principal Strains and Thickness Reduction in Single Point Incremental Forming." *Bull Series D* 74, no. 2 (2012): 111–120.
59. Arfa, H., R. Bahloul, and H. BelHadjSalah. "Finite Element Modelling and Experimental Investigation of Single Point Incremental Forming Process of Aluminum

Sheets: Influence of Process Parameters on Punch Force Monitoring and on Mechanical and Geometrical Quality of Parts." *International Journal of Material Forming* 6, no. 4 (2013): 483–510.

60. Blaga, Adrian, Octavian Bologa, Valentin Oleksik, and Radu Breaz. "Influence of Tool Path on Main Strains, Thickness Reduction and Forces in Single Point Incremental Forming Process." *Proceedings in Manufacturing Systems* 6, no. 4 (2011): 191–196.
61. Duflou, J.R., Bart Callebaut, Johan Verbert, and Hans De Baerdemaeker. "Improved SPIF Performance Through Dynamic Local Heating." *International Journal of Machine Tools and Manufacture* 48, no. 5 (2008): 543–549.
62. Mohammadi, Amirahmad, Ling Qin, Hans Vanhove, Marc Seefeldt, Albert Van Bael, and Joost R. Duflou. "Single Point Incremental Forming of an Aged Al-Cu-Mg Alloy: Influence of Pre-heat Treatment and Warm Forming." *Journal of Materials Engineering and Performance* 25, no. 6 (2016): 2478–2488.
63. Chera, Ionut, Octavian Bologa, Gabriel Racz, Radu Breaz, and Mihai Crenganis. "FEM Researches Regarding Incremental Forming Process." *Ann Oradea Univ Fascicle Manag Technol Eng* 1 (2013): 53–58.
64. Al-Obaidi, Amar, Verena Kräusel, and Dirk Landgrebe. "Hot Single-Point Incremental Forming Assisted by Induction Heating." *The International Journal of Advanced Manufacturing Technology* 82, no. 5–8 (2016): 1163–1171.
65. Asghar, Javed, and N. Venkata Reddy. "Importance of Tool Configuration in Incremental Sheet Metal Forming of Difficult to Form Materials Using Electro-Plasticity." In: *Proceedings of the World Congress on Engineering 2013*, Vol III, WCE 2013, July 3 - 5, 2013, London, U.K. (2013): 1734–1738.
66. Thibaud, S., R. BenHmida, F. Richard, and P. Malécot. "A Fully Parametric Toolbox for the Simulation of Single Point Incremental Sheet Forming Process: Numerical Feasibility and Experimental Validation." *Simulation Modelling Practice and Theory* 29 (2012): 32–43.
67. Al-Ghamdi, Khalid A., and G. Hussain. "Forming Forces in Incremental Forming of a Geometry with Corner Feature: Investigation into the Effect of Forming Parameters Using Response Surface Approach." *The International Journal of Advanced Manufacturing Technology* 76, no. 9–12 (2015): 2185–2197.
68. Al-Ghamdi, K.A., G. Hussain, and Shahid I. Butt. "Force Variations with Defects and a Force-Based Strategy to Control Defects in SPIF." *Materials and Manufacturing Processes* 29, no. 10 (2014): 1197–1204.
69. Bagudanch, I., M.L. Garcia-Romeu, G. Centeno, A. Elías-Zúñiga, and J. Ciurana. "Forming Force and Temperature Effects on Single Point Incremental Forming of Polyvinylchloride." *Journal of Materials Processing Technology* 219 (2015): 221–229.
70. Durante, M., A. Formisano, A. Langella, and F. Memola Capece Minutolo. "The Influence of Tool Rotation on an Incremental Forming Process." *Journal of Materials Processing Technology* 209, no. 9 (2009): 4621–4626.
71. Xu, Dongkai, Bin Lu, Tingting Cao, Jun Chen, Hui Long, and Jian Cao. "A Comparative Study on Process Potentials for Frictional Stir-and Electric Hot-Assisted Incremental Sheet Forming." *Procedia Engineering* 81 (2014): 2324–2329.
72. Honarpisheh, M., M.J. Abdolhoseini, and S. Amini. "Experimental and Numerical Investigation of the Hot Incremental Forming of Ti-6Al-4V Sheet Using Electrical Current." *The International Journal of Advanced Manufacturing Technology* 83, no. 9–12 (2016): 2027–2037.
73. Blaga, Adrian, and Valentin Oleksik. "A Study on the Influence of the Forming Strategy on the Main Strains, Thickness Reduction, and Forces in a Single Point Incremental Forming Process." *Advances in Materials Science and Engineering* (2013): 1–10.
74. Filice, L., G. Ambrogio, and F. Micari. "On-Line Control of Single Point Incremental Forming Operations Through Punch Force Monitoring." *CIRP Annals* 55, no. 1 (2006): 245–248.

# 5 Incremental Sheet Forming Parameters

## 5.1 INTRODUCTION

The features, quality and cost of manufactured components set the competitiveness and suitability of this die-less process. The parameters of a manufacturing process determine the quality of the components to be produced. In addition, the features and manufacturing cost of the components are directly related to the parameters of the manufacturing processes. In ISF, the parameters control the cycle time of the process that again in turn affects the manufacturing cost and productivity of the process. Furthermore, various responses of the ISF process like forming-force, formability, surface roughness, strain, dimensional accuracy and microstructure of the formed components are controlled and determined by the parameters of the ISF process. Significant developments and new solutions in the ISF process have demonstrated the importance of this process in various sectors like medical, automobile and aerospace. Nevertheless, this novel process has not been fully implemented in high-value industries due to a lack of available knowledge regarding the process parameters that limits the process for industrial applications. ISF suitability can be enhanced on the industrial scale when significant guidelines are highlighted regarding the relation between input parameters and responses of the process (forming-force, formability, surface roughness, strain, dimensional accuracy and microstructure of formed components, etc.). This chapter provides the technological capabilities and particular drawbacks of the process in terms of the ISF parameters.

## 5.2 CLASSIFICATIONS OF PARAMETERS AFFECTING ISF

ISF is influenced by different parameters. These parameters affect the ISF process and are classified into three major groups: Workpiece parameters, machine parameters and process parameters, as shown in Figure 5.1. The workpiece parameters are further divided into two categories as material parameters and geometrical parameters. The parameters in each group are also shown in Figure 5.1 and their influence on the successful and economic forming of the components is explained in the following sections.

## 5.3 WORKPIECE PARAMETERS

Workpiece parameters are related to the geometry and material of the component to be formed using the ISF process, as shown in Figure 5.1.

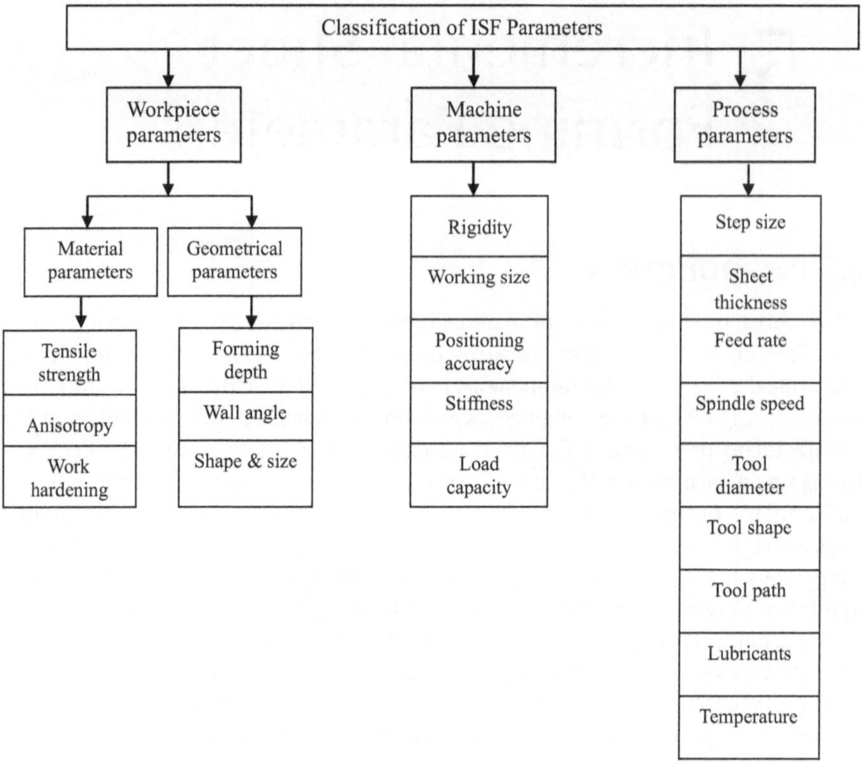


**FIGURE 5.1** Classification of ISF parameters.

### 5.3.1 Material Parameters

Material parameters are those factors which are related to the component material's characteristics and affect the ISF process significantly. Important material parameters that are significant in the ISF process are:

- Tensile strength
- Anisotropy
- Work-hardening exponent

These material parameters can be easily determined by uni-axial tensile tests. These parameters change with the change in materials [1]. Various materials have been investigated by different researchers as shown in Tables 4.1, 4.2 and 4.3 (see Chapter 4). Aerens et al. [2] showed that the tensile strength of materials affects the required forming-forces significantly. They investigated different materials of different tensile strengths and found that the forming-force (axial force) increases with the increase in tensile strength, as shown in Figure 5.2. Ajay [1] also investigated the impact of tensile strength, strain hardening exponent and strength coefficient on forming-force, formability, strain and surface roughness. Results showed that the

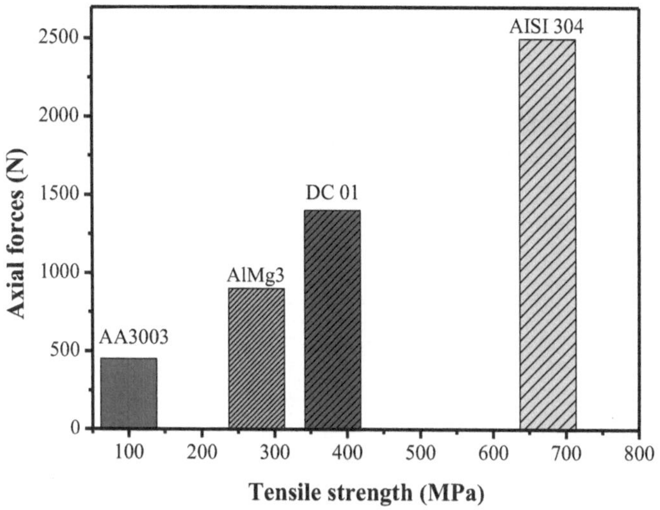


**FIGURE 5.2** Relation between axial force and tensile strength for different materials.

**TABLE 5.1**
**Material Parameters Investigated in ISF by Ajay [1]**

| Material | Sheet Thickness (mm) | Tensile Strength (MPa) | Strain Hardening Exponent | Strength Coefficient (MPa) | Summary |
|---|---|---|---|---|---|
| AA2024-O | 0.8 | 220 | 0.17 | 390 | Forming-force, strain, surface roughness and formability increased with the increase in tensile strength, strain hardening exponent and strength coefficient |
| | 1.0 | 231 | 0.19 | 398 | |

forming-force, strain, surface roughness and formability increased with the increase in tensile strength, strain hardening exponent and strength coefficient. Table 5.1 shows the parameters investigated by Ajay [1]. Kumar et al. [3] also investigated the impacts of tensile strength, strain hardening exponent and strength coefficient on the formability of AA2024-O and confirmed a similar trend of results.

### 5.3.2 Geometrical Parameters

Geometrical parameters are factors related to the geometry of the component to be formed and influence the ISF process significantly. The forming depth, forming shape and wall angle are three important geometrical parameters that influence the ISF process, and these are shown in Figure 5.3 along with other ISF process

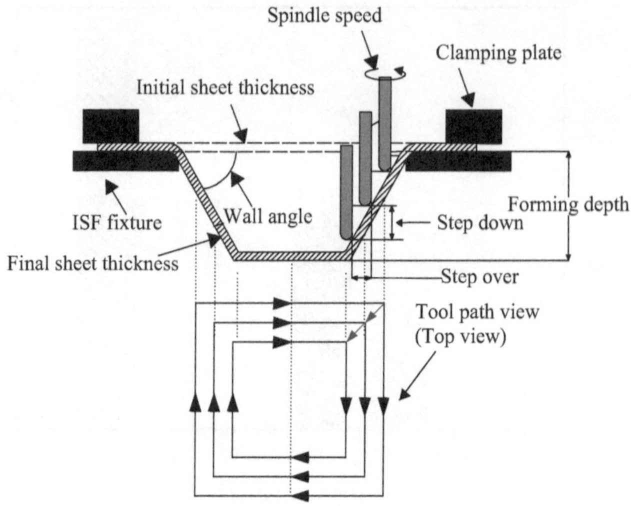


**FIGURE 5.3** Illustration of some process parameters of the ISF process.

parameters. The effect of geometrical parameters, as indicators of formability, is described in detail in Chapter 6.

*Forming depth*: It is the total depth or height of the component to be formed during the ISF process as shown in Figure 5.3. The total depth of the component is decided in the design phase by the requirements of the component. The forming depth is an important parameter for the ISF process which determines whether the component can be produced successfully or not by ISF. It is a well-known fact that the sheet material fails for the constant wall angle if the depth of the component increases beyond limits. Hence, this parameter can also be taken as an indicator of the formability of the components in ISF [3]. The depth achieved at the fracture can be considered as the measure of formability. This parameter also depends upon other factors like wall angle, step size, tool diameter, sheet thickness, tool-path, etc. For instance, with a thicker workpiece sheet a larger forming depth of the components can be produced keeping all other parameters constant [1].

*Forming shape*: It is the shape of the component to be formed. The shape and size of the components are crucial parameters that determine the applicability, reliability and suitability of the ISF process. A designer determines these parameters based on several factors like the need of customers, feasibility of the process and cost of product, etc. The effects of the shape and size of the components on ISF are complex and can be understood for different responses and outputs.

For instance, forming-forces and formability are two major responses in the ISF process which determine the safety of the forming machinery and forming-tools, and the accurate and safe forming of components. Since deformation occurs locally during the ISF process, the effects of the shape of the component on the forming-forces should be negligible. Duflou et al. [4] studied the impact of pyramidal and conical shapes with different wall angles on AA3103-O sheets and found that the forming-forces were identical for both shapes. Wang et al. [5] also formed pyramidal and conical frustums and showed that the forming-forces were independent of the part geometry. Arfa et al. [6] performed experimental work on AA3003-O sheets and analyzed the impact of pyramidal and conical shapes on the required forming-forces during the SPIF process. They also reported that the forming-forces were independent of the part geometry.

*Wall angle*: Wall angle is also a geometrical feature of the component to be produced and is the angle between the horizontal un-deformed and deformed sheet metal as shown in Figure 5.3. The wall angle is a key parameter of the ISF process which determines the formability of the material and the success of the ISF process. It is also considered an indicator of formability. Some researchers have determined the formability in terms of the limiting wall angle [3]. The limiting wall angle (i.e. the maximum wall angle that can be formed without fracture of the sheet material) is normally determined with the help of the cone forming test, as described in Section 3.2.2 in Chapter 3.

On the other hand, the shape and size of the components also affect the formability of the material due to the change in the mechanism of sheet-thinning. There is less sheet-thinning in constant wall angle frustums as compared to in varying wall angle objects [7]. As a matter of fact, a material fails due to excessive thinning of the sheet material during the ISF process which puts a barrier on the safe forming shape of the components. Material failure can be controlled by taking the constant wall angle of objects into account in order to achieve the enhanced formability. Lu et al. [8] also showed that the formability is greater for conical frustums as compared to that of pyramidal frustums for the same wall angle. Various shapes have been investigated for the ISF process, for different materials as shown in Tables 4.2 and 4.3 (in Chapter 4).

## 5.4 MACHINE PARAMETERS

Machine parameters are the parameters which affect the performance of the forming machinery and influence the ISF process significantly. The impacts of various ISF machine parameters on the ISF process have been discussed in detail in Chapter 4, Section 4.2. The requirements of ISF machinery and ISF tools are also directly related to the process parameters, discussed next.

## 5.5 PROCESS PARAMETERS

The third category of ISF parameters is the process parameters of the ISF process that play the most crucial role in executing the ISF process safely and successfully. The requirements of the forming machinery and forming-tools are also directly related to the process parameters of this die-less process [9]. Moreover, the ISF process parameters provide flexibility to the ISF process designers and engineers in controlling the ISF process [10, 11]. The different process parameters (see Figure 5.1) and their influences are discussed below.

### 5.5.1 Sheet Thickness

Sheet thickness is also a crucial process parameter of the ISF process. It is the initial thickness of the sheet to be formed into the desired shape. Researchers have used sheets of different thicknesses and different materials ranging from 0.5 mm to 2.0 mm during the ISF process. Sheet thickness determines the forming-force required, formability, dimensional accuracy and surface roughness of the formed components significantly. For instance, a larger thickness requires a greater forming-force to produce plastic deformation of the material. A higher thickness gives higher formability of the material. According to the sine law (described in Chapter 3), the thickness of the sheet decreases during the forming process as the wall angle increases; hence at the critical wall angle, the sheet material fractures due to the fact that the thinned sheet material is not able to resist the applied force. Hence, excessive thinning of the sheet leads to failure of the sheet material. Sheet thickness reduction, therefore, can also be considered as an indicator of failure [1].

### 5.5.2 Step Size

It is the size or distance by which the forming-tool is pressed into the sheet. There are two step sizes – step down defined in a direction perpendicular to the un-deformed sheet and step over defined as perpendicular to step down, as shown in Figure 5.3. Step down size is the distance by which the forming-tool moves in a downward direction after each contour. Similarly, the step over size is the distance by which the forming-tool moves in the horizontal inwards direction, after each contour of the forming operation. Step size is the most crucial ISF process parameter that determines the forming-force, formability, dimensional accuracy, surface roughness and quality of formed components, and the specification for the forming machinery and forming-tool.

### 5.5.3 Feed Rate

The feed rate is the motion of the tool relative to the sheet in unit time. It is the speed at which the forming-tool moves on the stationary sheet during the ISF process. The feed rate is also an important process parameter which determines the forming-time of components. A higher feed rate leads to less forming time or cycle time of components to be formed and vice-versa. The feed rate is normally described in mm/min and ranges from very low values (200 mm/min) to considerable values (5000 mm/min) in ISF. The effects of the feed rate on various responses are explained in Chapter 6 in detail.

### 5.5.4 Spindle Speed

The spindle speed is the speed at which the tool rotates at its axis during forming, and is an important process parameter of the ISF process. The forming-tool, mounted in the spindle of the ISF machine, can be rotated at different speeds as follows:

- Lock the spindle for rotation, i.e. zero spindle speed.
- Keep the spindle free for rotation, i.e. spindle rotates according to tool–sheet interaction.
- Spindle rotates with a specific spindle speed.

In the last case, the friction at the tool–sheet interface increases. An increase in the spindle speed increases the friction at tool–sheet interface which leads to an increase in the temperature of the sheet, leading to an increase in the ductility of the material that in turn requires lower forming-forces to deform the material [12]. Hence, smaller machinery can be used for executing the ISF process. The formability of the components is also increased at higher spindle speeds.

### 5.5.5 Geometry of the Forming-Tool

The forming-tool deforms the sheet material contour by contour; therefore, its shape and size (geometry) are important process parameters that impact the forming-force, formability, dimensional accuracy, thinning limit and microstructure of the component material significantly. The forming-tool geometry has been discussed in Chapter 4 (Section 4.3) in detail, and their effects on ISF are summarized in Table 4.1.

In conventional ISF, the forming-force increases with the increase in tool diameter as a result of a larger contact-zone between the tool and sheet. On the other hand, a larger tool diameter decreases the cycle time of the component significantly without compromising the surface quality of the part [13].

### 5.5.6 Tool-Path

The tool-path is the path or contour in which the forming-tool is moved to obtain the desired shape of the component, layer by layer, in the ISF process. The tool-path or contour is an important parameter of the ISF process that determines the dimensional accuracy of the component. Moreover, the tool-path also determines the forming-force required that in turn becomes useful in determining the requirements of the ISF machinery. In addition, the appropriate tool-path largely improves the formability, surface finish and dimensional accuracy of the formed component including the processing time [13]. For the ISF process, the tool-path can be obtained from CAM software that is available for manufacturing operations [14–16]. Tool-paths can be classified generally into two types, viz. the profile tool-path and the helical tool-path, as shown in Figure 5.4.

In the profile tool-path, the tool is given a step down and moves in a plane until it reaches its initial position; then it is again given a step over followed by a step down

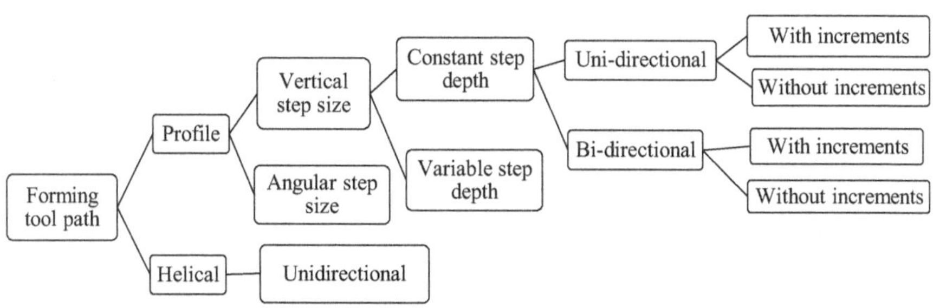


**FIGURE 5.4** Classification of tool-paths for ISF process.

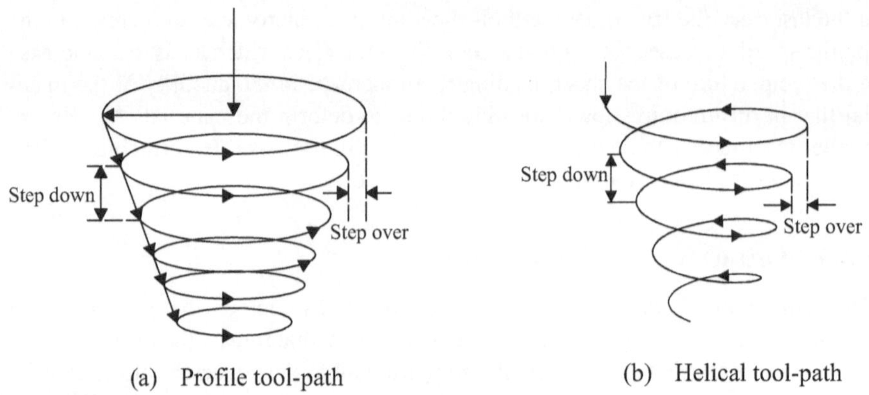


**FIGURE 5.5** Two possible tool-paths used in ISF for producing a truncated cone: (a) profile tool-path, (b) helical tool-path.

(see Figure 5.5a). This process continues until the complete shape is formed. The profile tool-path is also known as a constant Z-level tool-path. In the case of the helical tool-path, the tool moves in a helical path with pitch equal to step down, along the periphery of the finished component, maintaining the predetermined gradual helix along the vertical direction as shown in Figure 5.5b.

Some researchers [17, 18] studied the effects of the constant Z-level tool-path (profile tool-path) and helical tool-path and found that the helical tool-path formed the target geometry successfully, whereas the constant Z-level tool-path resulted in cracking of the sheet before obtaining the target depth. Moreover, the force exerted by the tool during the helical tool-path is more stable, and no unexpected force drop occurs in the vertical direction [6, 15].

The helical tool-path is also known as the "continuous tool-path", which provides stability to the forming-forces and produces better surface quality of the components. On the other hand, the profile tool-path is also known as the discontinuous tool-path, and produces stretch marks after each contour and results in instability of the force signals during the ISF process [6, 13, 15].

The profile tool-path can further be classified into the following two types:

- Straight step down
- Angular step down

In the straight step downtool-path (Figure 5.6a), the tool first moves inwards by required step over and then moves in the axial direction with constant step down after each contour. In the case of angular step down tool-path (Figure 5.6b), the tool moves in constant angular steps after each contour. Angular step down has not been found favorable as compared to vertical step down in the ISF process.

Straight step down can further be categorized into the following two types:

- Constant straight step down size
- Variable straight step down size or constant scallop height step size

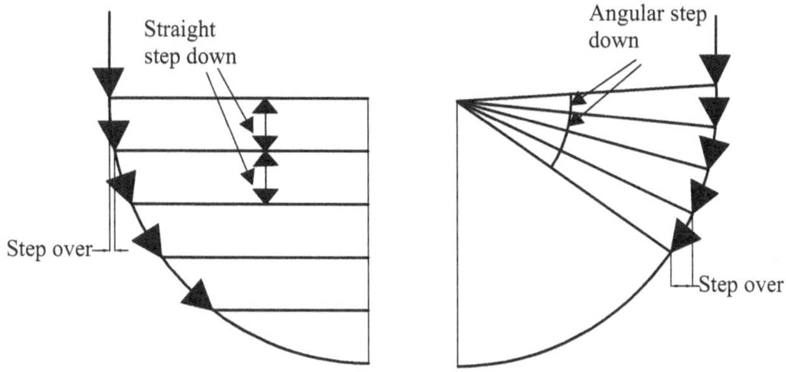


(a) Straight step down (b) Angular step down

**FIGURE 5.6** Profile tool-paths: (a) straight step down, (b) angular step down.

Constant straight step down is a step down path in which the forming-tool moves from the top to the bottom, contour by contour, with constant straight step down (Figure 5.7a). In this type of tool-path, the stretch marks are left at the transition point between consecutive contours and it produces poor surface quality with high values of step down (more than 0.5 mm). On the other hand, variable straight step down or constant scallop height step size is a step down path in which the forming-tool follows a series of consecutive contours with variable step down in order to keep the value of the scallop height constant (Figure 5.7b). This approach of moving the tool over the surface of the sheet avoids stretch marks and leads to better surface quality of the formed components [19].

The constant straight step down type tool-path can be executed in two ways as follows:

- Unidirectional
- Bidirectional

The unidirectional tool-path is the general form of the profile tool-path as shown in Figure 5.8a (for a pyramid-shape component). In this tool-path, the tool moves in a single plane and reaches its initial point, then it takes a step down in the straight direction before completing the next contour along the same direction. The unidirectional profile tool-path produces scratch marks along a line at which tool takes step down size. It also leads to twist in the workpiece. To avoid twist in the workpiece, a bidirectional profile tool-path (Figure 5.8b) can be executed which is similar to the profile tool-path except that in each subsequent cycle the tool changes the direction of motion and minimizes the twist and enhances the geometric accuracy of the formed component [20].

The unidirectional and bidirectional profile tool-paths can be further executed using two approaches as follows:

- With distributed increments
- Without distributed increments

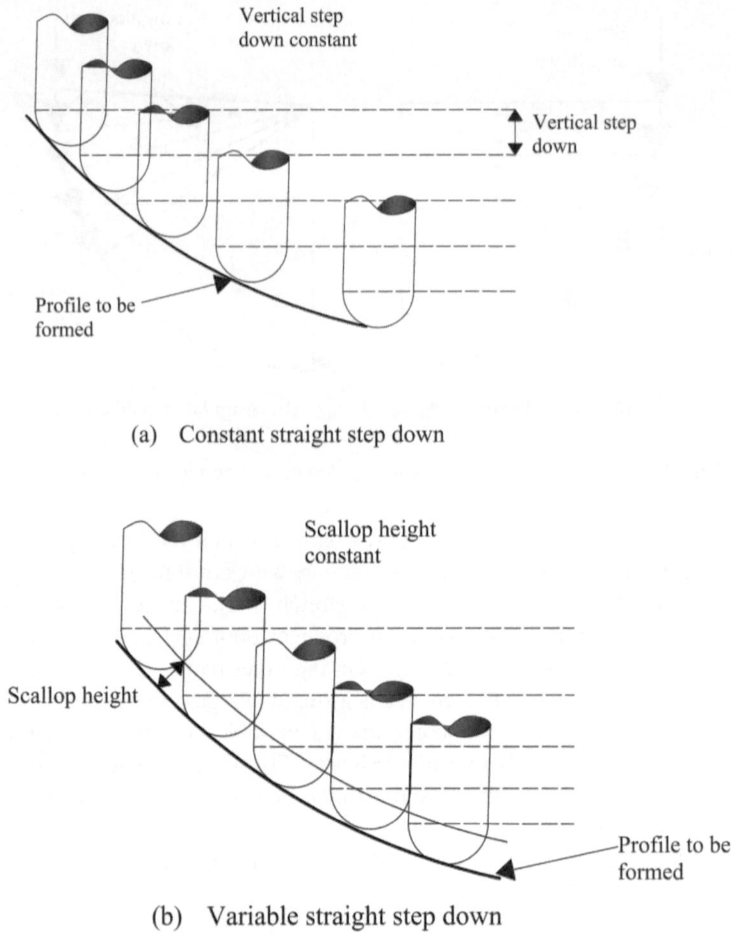


**FIGURE 5.7** Tool-path: (a) with constant straight step down, (b) with variable straight step down.

The unidirectional and bidirectional profile tool-paths can be executed with or without distributed increments. In the case of distributed increments, the tool completes a cycle and a quarter of the tool-path cycle is added before it moves to the next cycle, and by doing this scar marks are not visible on the surface. Also forming-forces are distributed uniformly along the edge of the geometry and the geometric accuracy of the part is enhanced [20]. These types of tool-paths are not normally available in commercial CAM packages. Figure 5.9 represents unidirectional and bidirectional profile tool-paths with distributed increments.

Wu et al. [21] proposed another way of distributed increment such that after forming each layer, the tool moves downward along the diagonal direction of the truncated pyramid wall surface in order to avoid local deformation in the corner area. Distributed increments in the diagonal direction are represented in Figure 5.10a and Figure 5.10b for unidirectional and bidirectional profile tool-paths respectively.

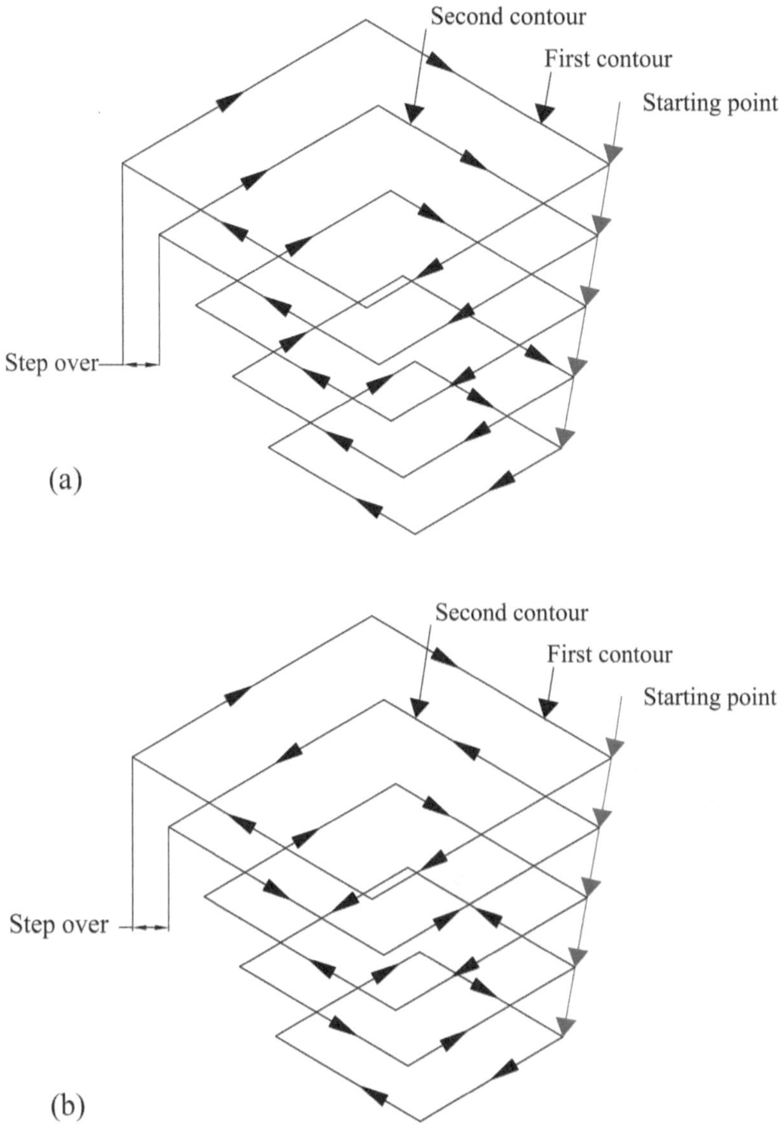


**FIGURE 5.8** Type of constant vertical step size type tool-path: (a) unidirectional, (b) bidirectional.

In addition, some researchers [22–37] have proposed different algorithms and approaches to tool-paths as summarized in Table 5.2.

### 5.5.7 Lubricants

The correct choice of lubricants determines the quality of components and the feasibility of the ISF process. This has been studied in detail in Chapter 4 (Section 4.5).

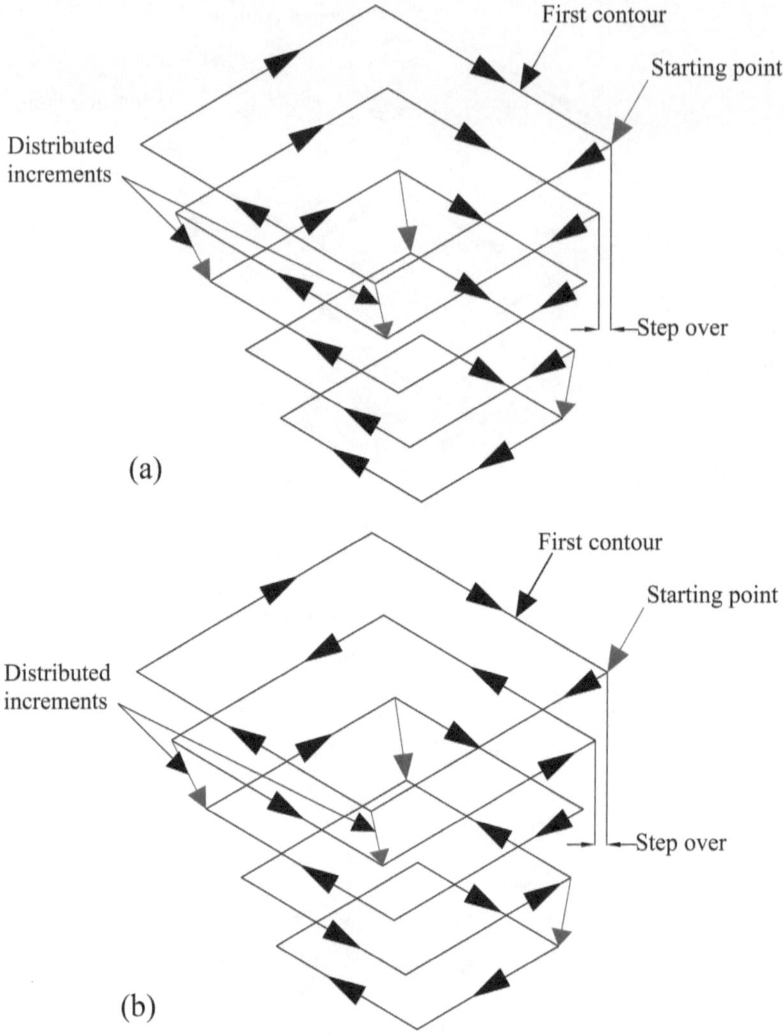


**FIGURE 5.9** Tool-paths with linear distributed increments: (a) unidirectional, (b) bidirectional.

### 5.5.8 Temperature

The temperature of the sheet is a critical parameter of the ISF process that impacts the forming-force, formability, dimensional accuracy, surface morphology, stress and strains developed in the component. The impact of temperature on ISF has been discussed in detail in Chapter 3 (Section 3.4).

## 5.6 CONCLUDING REMARKS

The ISF process has not been fully exploited in high-value industries due to the lack of available knowledge regarding the process parameters that control the process

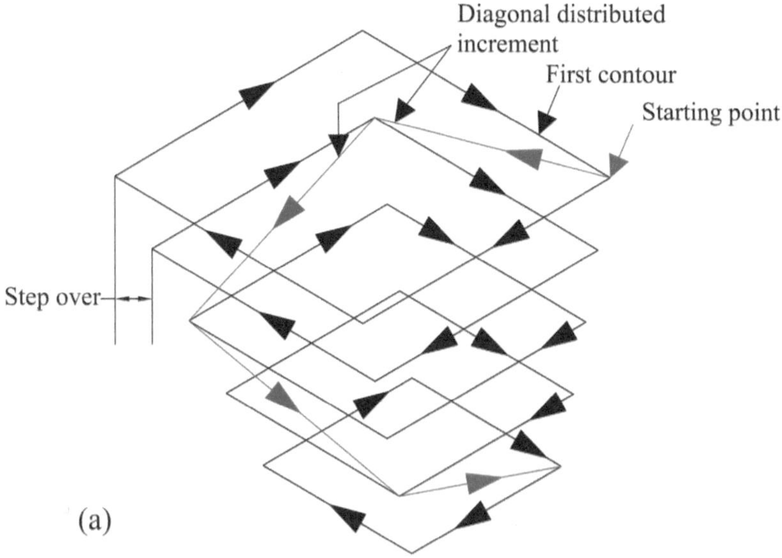


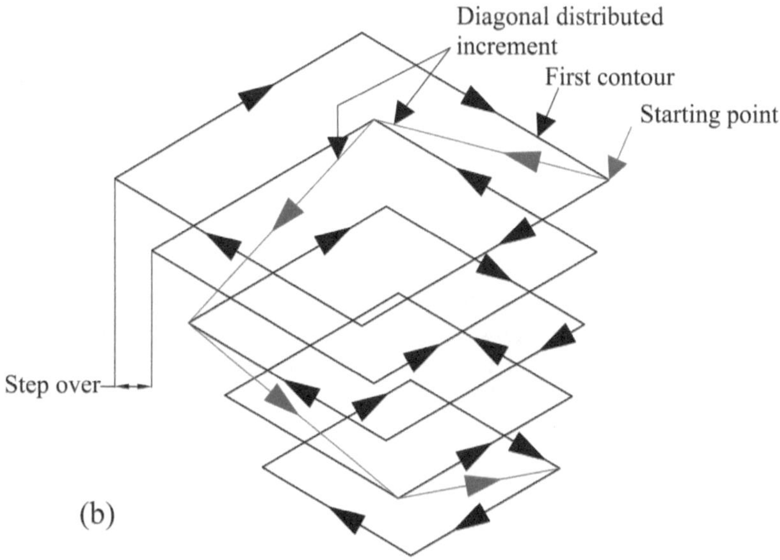


**FIGURE 5.10** Tool-paths with diagonal distributed increments: (a) unidirectional, (b) bidirectional.

and limit its industrial applications. ISF suitability can be enhanced on the industrial scale if significant guidelines are available regarding the relation between ISF input parameters and responses. ISF parameters are categorized into workpiece parameters, machine parameters and process parameters. However, the workpiece and machine parameters are critical but cannot be modified due to several constraints. Therefore, an ISF process designer has the choice of controlling the process parameters for forming the components successfully and economically using

**TABLE 5.2**
**Summary of Various Tool-Path Strategies Proposed by Researchers**

| Reference | Method(s) |
|---|---|
| [22, 23] | Feature-based tool-path correction |
| [24] | Tool-path compensation using mirroring |
| [25] | Online tool-path correction |
| [26] | Tool-path correction algorithms for deep geometries |
| [27, 28] | Tool-path strategy for back-drawing |
| [29] | Algorithms to convert the profile tool-path into helical tool-path |
| [30] | Tool-path strategy for multistep forming |
| [31] | Tool-path algorithm for kinematic tool and bottom forming |
| [32] | Squeezed and mixed tool-path |
| [33] | Spiral tool-path generation |
| [34] | Tool-path based on intelligent sequencing and multi-step mesh morphing |
| [35] | Tool-path using multivariate adaptive regression splines |
| [36] | Feature-based tool-path using graph topology |
| [37] | Tool-path based on feature detection algorithm |

ISF. Workpiece parameters relate to material parameters like the tensile strength, anisotropy and work-hardening exponent, and geometrical parameters like the forming depth, shape and wall angle of components. The wall angle is a key factor of the ISF process that can determine and control various outcomes of the process. Furthermore, the process parameters provide liberty to the process designers and engineers to a larger extent to control the process. The tool-path is, undoubtedly, an important parameter and feature of the ISF process which has the potential to control the dimensional accuracy, surface finish and aesthetic properties of the components. In this chapter, various tool-path strategies have been elaborated with their potential uses that would help the researchers and readers to understand the state-of-the-art techniques of performing the ISF process. The next chapter focuses on methods of measuring various responses occurring in the ISF process and the effects of parameters on the same.

## REFERENCES


1. Ajay. "Optimization of Process Parameters in Incremental Sheet Forming: Experimental and Theoretical Analysis." *PhD Thesis*, Guru Jambheshwar University of Science and Technology, Hisar, India (2018).
2. Aerens, Richard, Philip Eycke ns, Albert Van Bael, and J.R. Duflou. "Force Prediction for Single Point Incremental Forming Deduced from Experimental and FEM Observations." *The International Journal of Advanced Manufacturing Technology* 46, no. 9–12 (2010): 969–982.
3. Kumar, Ajay, Vishal Gulati, Parveen Kumar, Vinay Singh, Brijesh Kumar, and Hari Singh. "Parametric Effects on Formability of AA2024-O Aluminum Alloy Sheets in Single Point Incremental Forming." *Journal of Materials Research and Technology* 8, no. 1 (2019): 1461–1469.

4. Duflou, Joost, Yasemin Tunckol, Alex Szekeres, and Paul Vanherck. “Experimental Study on Force Measurements for Single Point Incremental Forming.” *Journal of Materials Processing Technology* 189, no. 1–3 (2007): 65–72.
5. Wang, Jyhwen, Mahesh Nair, and Ying Zhang. “An Efficient Force Prediction Strategy in Single Point Incremental Sheet Forming.” *Procedia Manufacturing* 5 (2016): 761–771.
6. Arfa, H., R. Bahloul, and H. BelHadjSalah. “Finite Element Modelling and Experimental Investigation of Single Point Incremental Forming Process of Aluminum Sheets: Influence of Process Parameters on Punch Force Monitoring and on Mechanical and Geometrical Quality of Parts.” *International Journal of Material Forming* 6, no. 4 (2013): 483–510.
7. Suresh, Kurra, SreeDivya Bagade, and Srinivasa Prakash Regalla. “Deformation Behavior of Extra Deep Drawing Steel in Single-Point Incremental Forming.” *Materials and Manufacturing Processes* 30, no. 10 (2015): 1202–1209.
8. Lu, B., Y. Fang, D.K. Xu, J. Chen, S. Ai, H. Long, H. Ou, and Jian Cao. “Investigation of Material Deformation Mechanism in Double Side Incremental Sheet Forming.” *International Journal of Machine Tools and Manufacture* 93 (2015): 37–48.
9. Kumar, Ajay, Vishal Gulati, and Parveen Kumar. “Experimental Investigation of Forming Forces in Single Point Incremental Forming.” In: Shanker Kripa, Shankar Ravi, & Sindhwani Rahul (eds.), *Advances in Industrial and Production Engineering*, pp. 423–430. Springer, 2019.
10. Kumar, Ajay, and Vishal Gulati. “Experimental Investigation and Optimization of Surface Roughness in Negative Incremental Forming.” *Measurement* 131 (2019): 419–430.
11. Kumar, Ajay, Vishal Gulati, and Parveen Kumar. “Effects of Process Parameters on Surface Roughness in Incremental Sheet Forming.” *Materials Today: Proceedings* 5, no. 14 (2018): 28026–28032.
12. Kumar, Ajay, Vishal Gulati, and Parveen Kumar. “Investigation of Surface Roughness in Incremental Sheet Forming.” *Procedia Computer Science* 133 (2018): 1014–1020.
13. Kumar, Ajay, Vishal Gulati, Parveen Kumar, and Hari Singh. “Forming Force in Incremental Sheet Forming: A Comparative Analysis of the State of the Art.” *Journal of the Brazilian Society of Mechanical Sciences and Engineering* 41, no. 6 (2019): 251.
14. Gulati, Vishal, Kumar, A. “Investigation of Some Process Parameters on Forming Force in Single Point Incremental Forming.” *International Research Journal of Engineering and Technology* 4, no. 4 (2018): 784–791.
15. Kumar, Ajay, and Vishal Gulati. “Experimental Investigations and Optimization of Forming Force in Incremental Sheet Forming.” *Sādhanā* 43, no. 10 (2018): 159.
16. Kumar, Ajay, Vishal Gulati, and Parveen Kumar. “Investigation of Process Variables on Forming Forces in Incremental Sheet Forming.” *International Journal of Engineering and Technology* 10 (2018): 680–684.
17. Blaga, Adrian, and Valentin Oleksik. “A Study on the Influence of the Forming Strategy on the Main Strains, Thickness Reduction, and Forces in a Single Point Incremental Forming Process.” *Advances in Materials Science and Engineering* (2013): 1–10.
18. Thibaud, Sébastien, R. Ben Hmida, Fabrice Richard, and Pierrick Malécot. “A Fully Parametric Toolbox for the Simulation of Single Point Incremental Sheet Forming Process: Numerical Feasibility and Experimental Validation.” *Simulation Modelling Practice and Theory* 29 (2012): 32–43.
19. Attanasio, Aldo, Elisabetta Ceretti, and Claudio Giardini. “Optimization of Tool Path in Two Points Incremental Forming.” *Journal of Materials Processing Technology* 177, no. 1–3 (2006): 409–412.
20. Suresh, Kurra, Arman Khan, and Srinivasa Prakash Regalla. “Tool Path Definition for Numerical Simulation of Single Point Incremental Forming.” *Procedia Engineering* 64 (2013): 536–545.

21. Wu, S.H., Ana Reis, F.M. Andrade Pires, Abel D. Santos, and A. Barata da Rocha. "Study of Tool Trajectory in Incremental Forming." *Advanced Materials Research* 472 (2012): 1586–1591.
22. Verbert, J. "Computer Aided Process Planning for Rapid Prototyping with Incremental Sheet Forming Techniques." *PhD Thesis*, Katholieke Universiteit Leuven, Leuven (2010).
23. Lu, B., J. Chen, H. Ou, and Jian Cao. "Feature-Based Tool Path Generation Approach for Incremental Sheet Forming Process." *Journal of Materials Processing Technology* 213, no. 7 (2013): 1221–1233.
24. Bambach, Markus, B. Taleb Araghi, and Gerhard Hirt. "Strategies to Improve the Geometric Accuracy in Asymmetric Single Point Incremental Forming." *Production Engineering* 3, no. 2 (2009): 145–156.
25. Rauch, Matthieu, Jean-Yves Hascoet, Jean-Christophe Hamann, and Yannick Plenel. "Tool Path Programming Optimization for Incremental Sheet Forming Applications." *Computer-Aided Design* 41, no. 12 (2009): 877–885.
26. Ambrogio, G., L. Filice, and G.L. Manco. "Considerations on the Incremental Forming of Deep Geometries." *International Journal of Material Forming* 1, no. 1 (2008): 1143–1146.
27. Ambrogio, G., L. De Napoli, and L. Filice. "A Novel Approach Based on Multiple Back-Drawing Incremental Forming to Reduce Geometry Deviation." *International Journal of Material Forming* 2, no. 1 (2009): 9–12.
28. Ambrogio, Giuseppina, and Luigino Filice. "On the Use of Back-Drawing Incremental Forming (BIF) to Improve Geometrical Accuracy in Sheet Metal Parts." *International Journal Of Material Forming* 5, no. 4 (2012): 269–274.
29. Skjødt, Martin, Michael H. Hancock, and Niels Bay. "Creating Helical Tool Paths for Single Point Incremental Forming." *Key Engineering Materials* 344 (2007): 583–590.
30. Skjødt, Martin, Niels Bay, Benny Endelt, and G. Ingarao. "Multi Stage Strategies for Single Point Incremental Forming of a Cup." *International Journal of Material Forming* 1, no. 1 (2008): 1199–1202.
31. Essa, Khamis, and Peter Hartley. "An Assessment of Various Process Strategies for Improving Precision in Single Point Incremental Forming." *International Journal of Material Forming* 4, no. 4 (2011): 401–412.
32. Malhotra, Rajiv, Jian Cao, Feng Ren, Vijitha Kiridena, Z. Cedric Xia, and N.V. Reddy. "Improvement of Geometric Accuracy in Incremental Forming by Using A Squeezing Toolpath Strategy with Two Forming Tools." *Journal of Manufacturing Science and Engineering* 133, no. 6 (2011): 1–10.
33. Malhotra, Rajiv, N.V. Reddy, and Jian Cao. "Automatic 3D Spiral Toolpath Generation for Single Point Incremental Forming." *Journal of Manufacturing Science and Engineering* 132, no. 6 (2010): 061003.
34. Behera, Amar Kumar, Bert Lauwers, and Joost R. Duflou. "Tool Path Generation for Single Point Incremental Forming Using Intelligent Sequencing and Multi-Step Mesh Morphing Techniques." *Key Engineering Materials* 554 (2013): 1408–1418.
35. Behera, Amar Kumar, Johan Verbert, Bert Lauwers, and Joost R. Duflou. "Tool Path Compensation Strategies for Single Point Incremental Sheet Forming Using Multivariate Adaptive Regression Splines." *Computer-Aided Design* 45, no. 3 (2013): 575–590.
36. Behera, Amar Kumar, Bert Lauwers, and Joost R. Duflou. "An Integrated Approach to Accurate Part Manufacture in Single Point Incremental Forming Using Feature Based Graph Topology." *Key Engineering Materials* 504 (2012): 869–876.
37. Behera, Amar Kumar, Bert Lauwers, and Joost R. Duflou. "Advanced Feature Detection Algorithms for Incrementally Formed Sheet Metal Parts." *Transactions of Nonferrous Metals Society of China*22 (2012): s315–s322.

# 6 Responses of Incremental Sheet Forming Process

## 6.1 INTRODUCTION

The control of parameters for any manufacturing process has a great role in the output of the process and the quality of the produced components. In addition, the implementation of any manufacturing process on an industrial scale is largely dependent on the guidelines that highlight the relationship between input parameters and responses (output parameters). In the previous chapter, various input parameters that directly control the ISF process were studied in detail. In this chapter, various output parameters (or responses) of ISF will be studied in detail. The responses of the ISF process to the input parameters determine the criteria for setting up the values of parameters for various applications. Furthermore, a process engineer can alter the process parameters to control the various responses of the ISF process to make it suitable for specific applications. The methods of measuring the output parameters are also explained in this chapter to impart knowledge on how to measure output parameters to the readers and researchers. Once the effects of input parameters on the output parameters are known, the process can be optimized, and various process models can be developed for predicting the behavior of the ISF process under different scenarios.

Although there are many responses of the ISF process that have been investigated by researchers, the important responses that determine the suitability and sustainability of the ISF process on an industrial scale are:

- Forming-force
- Formability
- Geometrical accuracy
- Thickness reduction
- Surface roughness

## 6.2 FORMING-FORCE

The forming-force is the force required to deform the sheet locally at the tool–sheet interface. The Cartesian coordinate frame is chosen such that the $z$-axis is along the forming-tool axis and the unreformed sheet is in the $x$–$y$ plane, as shown in Figure 6.1. The next section gives the basic set-up and process of forming-force measurement and the effects of various ISF parameters on the forming-force are discussed next.

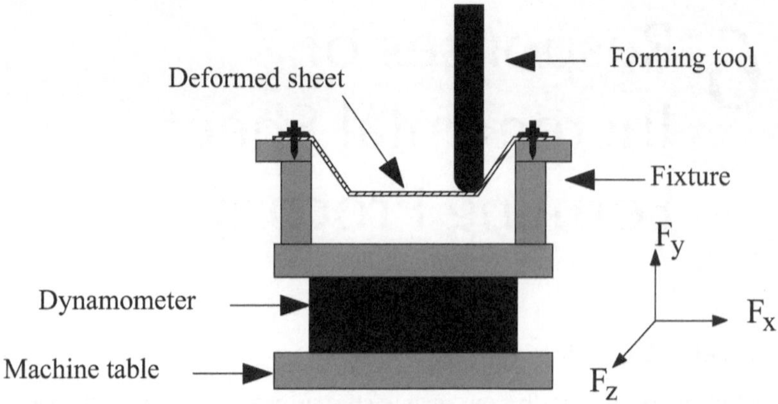


**FIGURE 6.1** Schematic of the measurement of components of forming-force using a force dynamometer.

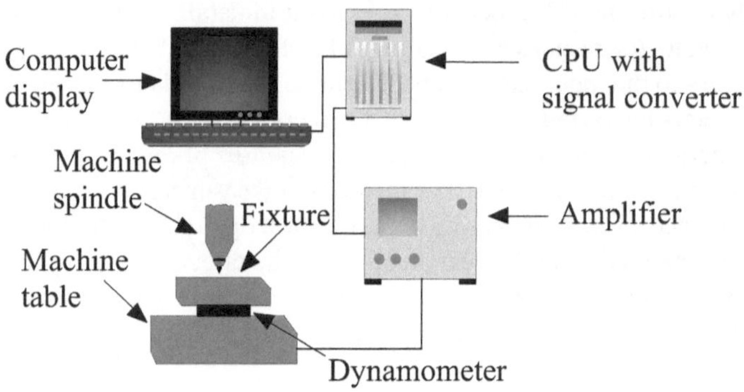


**FIGURE 6.2** Set-up for the measurement of forming-forces in the ISF process.

### 6.2.1 Measurement of the Forming-Force

The force components *Fx*, *Fy* and *Fz* along the *x*, *y* and *z* directions, respectively, can be measured with the help of a force dynamometer or a load cell. In order to measure the forming-forces required to form the desired shape in the incremental forming process, the force dynamometer or load cell can be placed below the ISF fixture in which sheet is clamped, and the data can be recorded with the help of a data acquisition system, charge amplifier and PC-based environment as shown in Figure 6.2. The load cell can also be mounted on the forming-tool for measuring the force components [1].

A data acquisition system contains a hardware–software set-up which allows it to display measured forces on the screen and to record data to facilitate the processing of force values in a PC-based environment. This software generates Excel sheets and graphs with real-time values of the measured forming-forces.

### 6.2.2 Effects of ISF Parameters on Forming-Force

The input parameters of the ISF process significantly affect the forming-force required to produce local deformation in the sheet during the forming operation. It is customary to investigate the forming-force in the ISF process for a proper understanding of the deformation mechanism, predicting the failure of the material, monitoring the forming process and online control of the process for successful ISF [2]. There are several input parameters in the ISF process that affect the forming-force during the forming operation. These are discussed in the next sections.

*Effects of tool path on the forming-force*: The forming-tool path is an important parameter of the ISF process that affects the forming-force in ISF, as explained in Chapter 5. Different types of tool-paths have been proposed and investigated by various researchers. Table 6.1 summarizes the results of various tool-paths on forming-forces, as reported by researchers. It can be concluded from Table 6.1 that the helical tool-path results in constant and lower forming-forces during the ISF process. The angular step-down tool-path has been proven as the worst type of tool-path because it requires larger forces to form the components as compared to straight step down and the helical tool-path. Forming-forces are lower with the profile tool-path having distributed increments as compared to those with the profile tool-path having no distributed increment [1].

*Effects of the tool diameter on the forming-force*: The tool diameter also affects the forming-force significantly in the ISF process. The forming-force increases with the increase in tool diameter in the conventional ISF process. Some researchers [8, 9] have also investigated the impact of the tool diameter for PVC and found the same trend in the forming-force. Kumar et al. [10] investigated the impact of the tool diameter on the axial peak forming-forces for different sheet thicknesses and found that the forming-force increases with the forming-tool diameter irrespective of sheet thickness. Their results are shown in Figure 6.3.

In the case of electric-assisted ISF, the forming-force decreases with the rise in tool diameter [11]. This is because the current density increases at the tool–sheet interface due to the larger tool diameter and the higher current density increases the temperature at the localized zone which increases the local ductility of the material [1], making it easier to deform.

*Effects of the sheet thickness on the forming-force*: The sheet thickness directly impacts the forming-forces during the ISF process. The forming-force increases with the increase in sheet thickness because more material has to be deformed in a single loop for the thicker sheet [12]. The effect of sheet thickness on the peak forming-force is also seen in Figure 6.4, and it is observed that the increase in forming-force is non-linear in nature.

*Effects of the step size on the forming-force*: The step down size determines how much deformation in the sheet material is produced in one cycle. A smaller step down implies less deformation and less forming-force required. But a smaller step down will require more cycles and, hence, more time to complete the ISF of a given component; therefore, the step down size is also a significant parameter that affects the forming-force. The forming-force increases with the increase in step down size for metal as well as for PVC during all variants of the ISF process [1].

**TABLE 6.1**
**Effects of Tool-Path on Forming-Force**

| Sr. No. | Tool-Paths Used | Sheet Material | Workpiece Shape | Results | Reference |
|---|---|---|---|---|---|
| 1 | Vertical step depth, angular step depth and helical | DC04 steel | Dome | Maximum axial force was lower (768N) with the helical tool-path than the forces observed with the other two tool-paths (871 N for vertical step and 1265 N for angular step down) | [3] |
| 2 | Profile, profile with distributed increments, helical | DC04 steel | Conical frustum | Maximum axial forming-force was lower for the helical tool-path (811 N) as compared to those produced for press strategy with increments (928 N) and profile tool-path (1004 N) | [4] |
| 3 | Profile, helical with 1° helix angle | FPG copper | Pyramidal frustum | With the helical path, a safe geometry was obtained. For the profile tool-path, crack occurred before the desired depth | [5] |
| 4 | Profile, helical | AA3003-O | Pyramidal frustum and conical frustum | Forming-forces in the helical tool-path were less and steady as compared to those produced with the profile tool-path | [6] |
| 5 | Profile, helical | AA7075-O | Pyramidal frustum and conical frustum | Sudden peaks in axial forces were observed between two consecutive layers in the profile tool-path. In the helical tool-path, force variation was smooth. The maximum forming-forces with the helical tool-path were less than those observed with the profile tool-path. | [7] |

*Effects of the spindle speed on the forming-force*: The spindle speed or forming-tool rotational speed can be either zero, free or a specific value. It has been found that the forming-force decreases with the increase in spindle speed. This is because a higher spindle speed increases the friction at the tool–sheet interface resulting in more friction and heat generation, and an increase in ductility of the material. Hence, with a higher spindle speed, thicker sheets can be formed with reduced forming-forces [1, 13].

*Effects of the feed rate on the forming-force*: The feed rate is also an important factor which affects the forming-force significantly. The forming-force increases with the increase in the feed rate during all variants of the ISF process because the material has to be deformed in a relatively short period of time [1].

*Effects of the wall angle on the forming-force*: The wall angle is a crucial factor for the forming-force in the ISF process. The forming-force increases with the rise in the

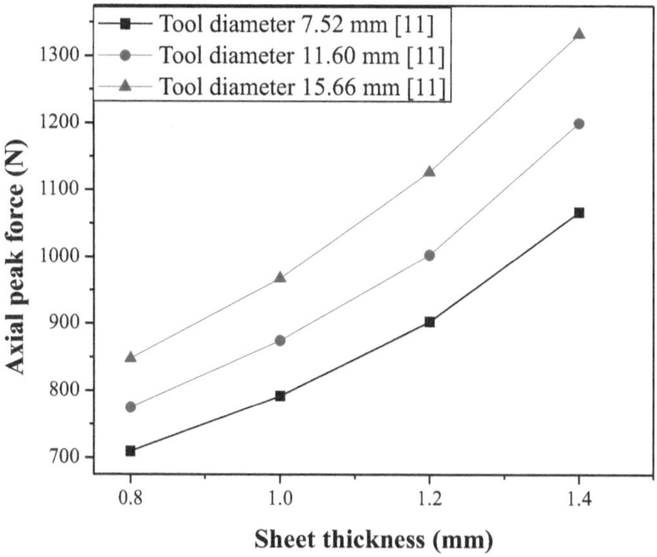


**FIGURE 6.3** Effects of sheet thickness and tool diameter on axial peak forces.

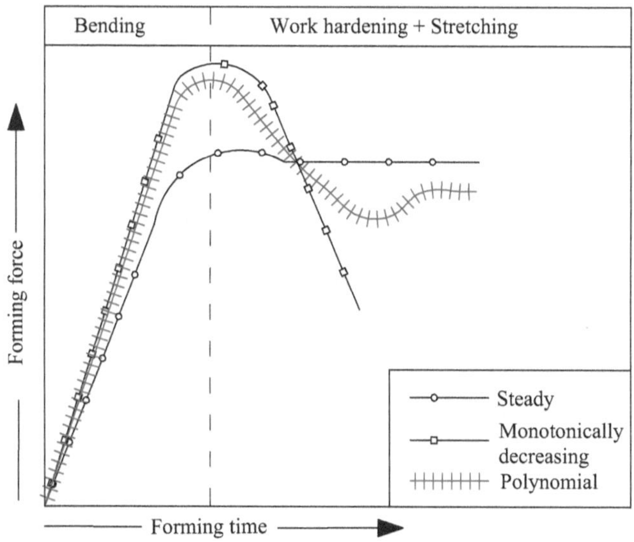


**FIGURE 6.4** Three trends of forming-force after peak value.

wall angle for metal and polymer sheets during the conventional ISF process (i.e. at room temperature). This is because the larger amount of material has to be deformed in a single contour for a higher wall angle [1, 13]. On the other hand, the effect of the wall angle is the opposite in electric-assisted ISF, i.e., forming-force decreases with the rise in wall angle. This happens because a larger wall angle increases the current density and the local temperature at the tool–sheet interface [1, 11].

*Effects of lubricants on the forming-force*: Lubrication is a key factor in the ISF process that determines the feasibility of forming the components safely using ISF. It not only improves tool life by dissipating heat from the forming-zone, it also removes abraded small particles of sheet material and/or forming-tool from the forming-zone. The effect of the use of lubricants in ISF has been discussed in detail in Chapter 4, Section 4.5. Several lubricants have been investigated by researchers for use in the ISF process. The results showed that the components can be formed without fracture when lubricants are used during the forming process. Failure of the material is observed at the early stage of forming when dry condition (without using any lubricant) is employed during the process. Severe tool wear is also observed during the dry forming of the components [1, 14, 15, 16].

*Effects of some other parameters on the forming-force*: In addition to various parameters discussed above, there are some other parameters also that impact the ISF process but have not been investigated conclusively, and require more investigation. One such parameter is workpiece geometry. It is well-known that the deformation is purely local during the ISF process. Hence, the shape of a component should not affect the forming-force [1]. Some researchers [15, 17, 18] have compared the forming-force for conical and pyramid frustums, and observed that the maximum axial forming-force is independent of these two-part geometries, while Al-Ghamdi et al. [19] showed that corners required 2.7 times higher forming-forces than straight walls. Pyramid-like shapes have corners and oblique walls whereas a conical shape has only oblique walls. During the forming of components with corners, a different forming-force requirement is expected. This is because oblique walls have less stiffness than corners. Hence, forming the corners of the pyramidal frustums should demand higher forming-forces as compared to those demanded by the oblique walls.

Another less explored factor is the tensile strength of the workpiece material. The forming-force increases with the increase in the tensile strength of the workpiece material [19, 20]. In ISF, the effects of sheet orientations are significant. The orientation of the sheet determines the strain-hardening exponent of the material [7]. A greater strain-hardening exponent of the material would increase the resisting capacity of the material to the plastic deformation, hence requiring a greater forming-force to deform the sheet material in the particular direction. The forming-force is also found to increase with the increase in the grain size of the material [21]. Table 6.2 gives the summary of the effects of different parameters discussed above on the forming-force during the ISF process.

### 6.2.3 Effects of the Deforming Mechanism on the Forming-Force

The deforming mechanism is greatly associated with the forming-force. Several researchers [22, 12, 23] have studied the impact of the deforming mechanism on the forming-force in the ISF process. The forming-force increases gradually up to a peak value due to bending mechanics in the initial stage of component shape forming. After reaching a peak value, the stretching mechanics take over, resulting in localized plastic deformation of the sheet. During the entire process a very complex force trend is observed, which is due to the combined effect of two different mechanisms induced in the material which are (i) work-hardening (results in the rise of forces)

**TABLE 6.2**
**Effects of Some Input Parameters on Forming-Force during the ISF Process**

| Name of Parameter | Effects on Forming-Force | |
|---|---|---|
| | **Conventional ISF** | **Electric-Assisted ISF** |
| **Tool diameter** | Increases with the increase in tool diameter | Decreases with the increase in tool diameter |
| **Sheet thickness** | Increases with the increase in sheet thickness | Increases with the increase in sheet thickness |
| **Step size** | Increases with the increase in step size | Increases with the increase in step size |
| **Spindle speed** | Decreases with the increase in spindle speed | Decreases with the increase in spindle speed |
| **Wall angle** | Increases with the increase in wall angle | Decreases with the increase in wall angle |
| **Presence of lubricant** | Decreases when lubricant is provided | – |
| **Tensile strength** | Increases with the increase in tensile strength | Increases with the increase in tensile strength |
| **Strain hardening exponent** | Increases with the increase in strain hardening exponent | – |

and (ii) sheet thinning (result of plastic deformation; contrarily this tends to reduce forces). Three different trends have been observed for forming-forces after reaching a peak value, namely

- Steady
- Polynomial
- Monotonically decreased

For a low value of wall angle (say 50° in SPIF), the forming-force becomes steady after reaching a peak value, and a kind of dynamic equilibrium is achieved because of the drop in the forming-force (due to material thinning) and the increase in force (due to strain-hardening). This trend for axial forces with a low wall angle remains independent of other parameters.

The polynomial trend of forming-forces is observed for severe but less than critical wall angles (say, between 50° and 60° in SPIF) and a negative gradient occurs right after the peak forming-force, which can be an influence of higher sheet thinning, and hence the forming-forces drop after the peak value. After this stage, the force trend becomes steady as an effect of equilibrium between work-hardening and sheet-thinning. For a higher value of wall angle (say, more than 70° in SPIF), the material yields to fracture, and material hardening is not able to compensate the influence of sheet thinning, resulting in the instability of the deformation process.

Thus, after the peak forming-force, a monotonically decreasing trend of forming-forces is observed and the force decreases until the sheet fractures. These three trends are shown in Figure 6.4.

According to the described conditions, a type of threshold between the stable and unstable conditions can be detected and the force gradient after peak values becomes a crucial indicator for the ISF process for a different set of input factors. It is clear that a process is safe and stable when a safe set of parameters is employed (say, low wall angles). Otherwise, the process fails due to excessive thinning of the sheet (at large step down size or wall angle). Hence, the gradient of the force curve after peak values can be considered and utilized as a hidden variable for modifying process parameters by continuous comparison of the crucial value and instant value for safe utilization of the ISF operation.

## 6.3 FORMABILITY

Formability is a critical response of the ISF process that can determine the suitability and robustness of the process for industrial applications. The formability of the ISF process is much higher than that of conventional forming processes due to the nature of the mechanism of deformation in this die-less process. The rudimentary deformation and fracture mechanism plays a vital role in optimizing the ISF process. Therefore, proper knowledge of the formability of the ISF process sets the criteria to explore the accuracy and thinning limits of this process.

### 6.3.1 Measurement of Formability

The formability of the ISF process can be measured in terms of following:

a) Maximum forming depth
b) Maximum wall angle
c) Forming limit diagrams

(a) *Maximum forming depth*: Forming depth is an important parameter of the ISF process which also determines whether the component can be produced successfully by ISF or not. The depth measured at the fracture can be considered the maximum forming depth and is a measure of the formability of ISF. The forming depth can be easily measured using contact and non-contact methods. In the contact method, the maximum forming depth (depth at fracture of the sheet) can be measured using a Vernier height gauge. In the non-contact method, the forming height can be measured using a 3D scanner or other methods.
(b) *Maximum wall angle*: Wall angle is also a key parameter of the ISF process which determines the formability of the ISF process. The limiting wall angle (i.e. maximum wall angle that can be formed without fracture of the sheet material) is normally determined with the help of the cone forming test (as described in Section 3.2.2, Chapter 3). The maximum forming angle can be easily measured with a coordinate-measuring machine (CMM). The

**TABLE 6.3**
**Summary of Maximum Attained Wall Angle for Different Materials in the ISF Process**

| Workpiece Shape | Tool Radius (mm) | Sheet Thickness (mm) | Material | Maximum Formed Wall Angle | Reference |
|---|---|---|---|---|---|
| Conical frustum | 5 | 0.5 | 65Cr2 | 57° | [24] |
| Conical frustum | 6 | 1 | Brass | 40° | [25] |
| Conical frustum | 6 | 1 | Copper | 65° | [25] |
| Conical frustum | 5 | 1 | DC01 | 67° | [26] |
| Conical frustum | 6 | 1 | DDQ | 70° | [25] |
| VWACF | 8 | 1 | DP600 | 68° | [27] |
| VWACF | 8 | 1 | DP780 | 45° | [27] |
| VWACF | 8 | 1 | DP1000 | 39° | [27] |
| VWACF | 7.5 | 3 | Polyvinyl chloride | 76° | [28] |
| VWACF | 7.5 | 3 | Polyethylene | 81° | [27] |
| Conical frustum | 6 | 1 | HSS | 65° | [25] |
| Conical frustum | 5 | 0.6 | TiAl6V4 | 32° | [24] |
| Conical frustum | 5 | 1 | AA2024-T3 | 42° | [24] |
| Conical frustum | 7.5 | 0.8 | AA2024-O | 65° | [22] |
| Conical frustum | 7.5 | 1 | AA2024-O | 68° | [22] |
| Conical frustum | 5 | 0.85 | AA3003-O | 70° | [24] |
| Conical frustum | 5 | 2 | AA3003-O | 76° | [24] |
| Conical frustum | 5 | 1.5 | AA1050-O | 76° | [26] |
| Conical frustum | 5 | 1.25 | AA5182 | 64° | [24] |
| Conical frustum | 12.7 | 0.93 | AA5182 | 63° | [29] |

maximum forming angle depends on the type of material, thickness of the sheet material and other process factors like temperature, tool diameter, step size, etc. Table 6.3 gives a summary of the maximum attained wall angle for various materials under different conditions by different researchers.

(c) *Forming limit diagram*: Forming limit diagram (FLD) is another way to explain the formability of the ISF process. Several research studies have been conducted to construct FLD for the ISF process [7, 29, 30, 31, 32, 33]. FLD represents the intersection of the major true strain axis and fracture forming limit line. The forming limit of the conventional forming process like deep drawing is known as the forming limit curve (FLC), which can be defined as a correlation between the major strain$\varepsilon_1$ and minor strain $\varepsilon_2$ in the plane of the sheet at the onset of necking failure. These strain values are recorded from a series of experiments under different loading conditions, ranging from biaxial tension (stretch-forming) to equal tension and compression (deep drawing) [34]. Experimental studies [31, 32, 35] have shown that the forming limits of the ISF process are well beyond the forming limits of conventional forming processes (i.e. FLCs). The forming limit of the ISF process can also be termed

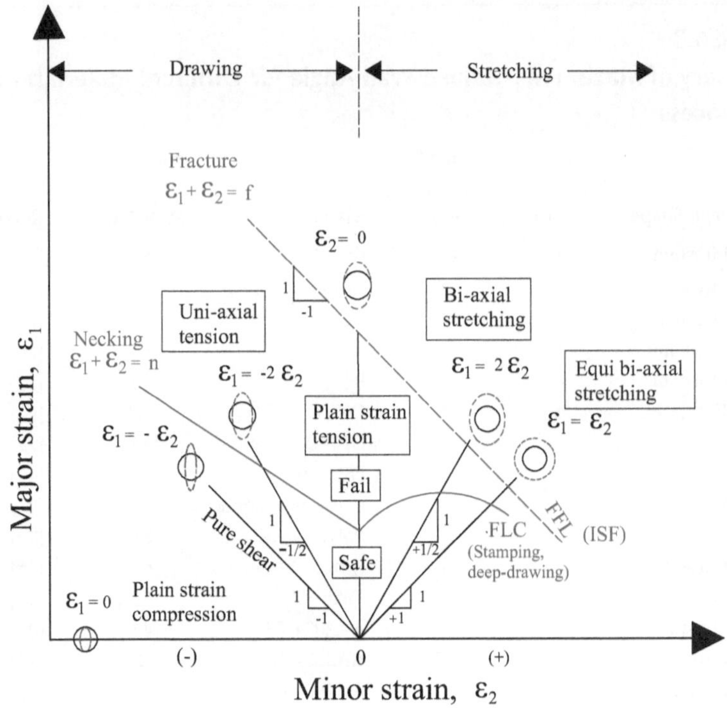


**FIGURE 6.5** Representation of FLD.

as the fracture forming limit (FFL). In the ISF process, sheet materials can be stretched much more than by the conventional forming processes. The FFL line is a straight line with a negative slope of unity in the positive major and minor strain region in FLD. This is the effect of deformation mechanics during the ISF process, i.e. plastic deformation under the forming-tool is purely localized resulting in achieving higher strains before fracture occurs. Figure 6.5 shows the FLD for the ISF process.

The ISF process experiences different non-uniform stresses and strains during the forming of the components. These non-uniform strains result in wrinkling and fracturing of material in the regions of high strains. It is important to spot these regions of high strains in order to avoid fracture by varying process parameters. Grid circle analysis (GCA) is a technique to spot the high and low strain regions on the surface of formed parts. GCA is one of the simple methods for analyzing local strains developed during the sheet material forming process. GCA can be performed using the following steps:

- Printing a grid of circles of definite diameter (generally 2 to 5 mm)
- Etching of printed circles
- Measurement of the major axis of ellipses formed during SPIF
- Calculation of true local strains

GCA consists of printing a grid of circles of specific diameter using the screen-printing method. Later, printed circles are etched into the material surface using a suitable technique so that the deformation of circles can be measured. Local forming of the sheet results in deformation of the circles which are etched on the sheet surface. Etched circles get converted into ellipses unless the strain is of a pure biaxial type. By measuring the major and minor diameters of the deformed circles, the strains can be determined for strain analysis. A CMM or video measuring machine (VMM) can be used to measure the major and minor axes diameters.

ISF is generally characterized by pure shear strains, and the formability of the process is increased by the uniform and local nature of forming with shear deformations [36]. As the curvature of the radius of the punch path rises, the strain mode converts into biaxial stretching (tension–tension) from plane-strain stretching (tension) (Figure 6.5) and the formability of the process is defined as ($\varepsilon_1 + \varepsilon_2$) where $\varepsilon_1$ and $\varepsilon_2$ are major and minor strains, respectively. Formability is higher in the plane strain condition ($\varepsilon_2 = 0$, as in conical frustum) and lower in biaxial stretching ($\varepsilon_2 \neq 0$, as in frustum with corners) [37, 38]. This specific shape of FFL permits the prediction of the forming ability of a material and process by the ratio $\varepsilon_2/\varepsilon_1$. The smaller the ratio is, the greater the formability. Hence, the formability of a material decreases with the increase of the ratio $\varepsilon_2/\varepsilon_1$. Minor strain is null when the process is pure shear, and, hence, formability would be maximum. It decreases when the process deviates from the condition of pure shear and minor strain increases [39].

### 6.3.2 Effects of Parameters on Formability

The input parameters of ISF greatly influence the formability of materials during the process. Table 6.4 shows the trends of the effects of process parameters on the formability of the ISF process.

## 6.4 GEOMETRICAL ACCURACY

Geometrical accuracy is a response of great importance in the ISF process that should be focused on and understood properly to enhance the viability of this dieless process on an industrial scale. In ISF, the forming of components in the desired shape is purely governed by the tool-path information based on component geometry. The forming-tool attempts to produce the target shape on the blank following the tool-path instructions. Hence, in ISF, the desired shape would only be obtained accurately if the produced strains in the sheet are purely in the plastic deformation range. The elastic property of the sheet material results in a local springback effect as the forming-tool moves on the sheet. Moreover, the development of residual stress and strain hardening is also expected because of the cyclic application of forming-force at the tool–sheet interface and the plastification of sheet metal, which is an inherent characteristic of the ISF process. A specific local region is generally subjected to tool contact from 10 to 100 times depending upon the tool diameter and step size. Therefore, the actual shape of the formed component after unloading from fixture is different from the desired CAD model due to elastic strains causing a springback effect [47, 48].

**TABLE 6.4**
**Summary of Effects of Input Parameters on Formability of the ISF Process**

| Parameter | Effects of Parameter on Formability | References |
|---|---|---|
| Temperature | Increases with the increase in temperature | [40, 41, 42] |
| Tool diameter | Decreases with the increase in tool diameter | [43, 44] |
| Sheet thickness | Increases with the increase in sheet thickness | [45] |
| Step size | Decreases with the increase in step size | [43, 45] |
| Spindle speed | Increases with the increase in spindle speed | [43, 44, 45] |
| Feed rate | Decreases with the increase in feed rate | [43, 44] |
| Wall angle | Decreases with the increase in wall angle | [45] |
| Tool shape | ORB tool produces higher formability than produced by hemispherical or vertical roller ball tool | [46] |
| Workpiece shape | Higher formability can be achieved with conical shape (plain strain condition) as compared to that of pyramidal shape (bi-axial stretching) | [30] |

It is a well-known fact that some tolerances are always given on the dimensions of a designed component for variations in the manufacturing process, material properties, etc. In a typical ISF process (say SPIF), the sheet is not supported from the bottom side and is only clamped along the periphery. For a benchmark shape, say truncated cone for which the springback effect is minimal, some geometrical error can be detected in the final shape after unloading as shown in Figure 6.6. The cause of the deviations from the desired shape can be [49]:

- Sheet bending near the major diameter of cone
- Wall deviation
- Pillow effect

Sheet bending near the major diameter of the cone shape is largely caused by the distance between the clamping edge and the forming area. By minimizing this gap and using a backing plate, the sheet bending effect can be reduced. The stiffness of the sheet can be increased by a clamping action and undesired movement of the blank

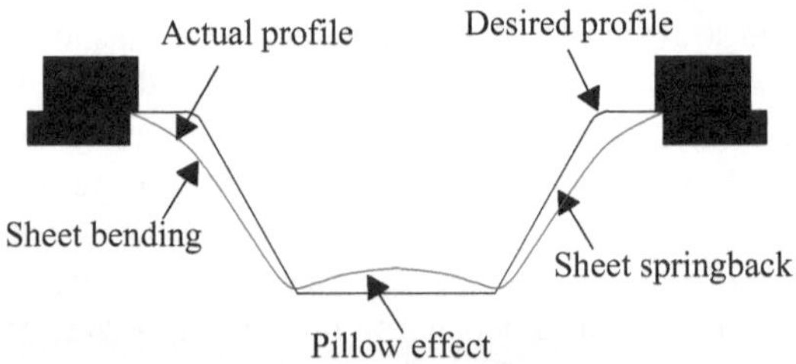


**FIGURE 6.6** Geometrical errors possible in the SPIF process.

can be decreased [50]. However, the use of a backing plate decreases the flexibility of the process and adds no benefit beyond the initial stages of the process [48]. The sheet material always lifts up after each contour of the tool trajectory and the final depth of the component becomes less than the target depth. The pillow effect is the concave shape formed at the minor diameter (base) of the cone. Dimensional error between the target shape and the actual formed shape can be reduced using optimum process parameters and tool-path [49]. A prediction of the amount of springback in formed parts is the key to the accurate prediction of the ISF process. Process accuracy can be enhanced by reducing springback effects and deviations from target shape [51].

The following are some approaches used to increase the geometrical accuracy of components formed using the ISF process:

- Using HA-ISF processes
- Online control of dimensional accuracy
- Using partial dies or specified forming punch
- Controlling the overlapping between successive contours
- Optimizing tool-paths
- Optimizing process parameters

### 6.4.1 Measurement of Geometrical Accuracy

Geometrical accuracy can be measured in terms of deviations of the dimensions of final component from the desired (or designed) dimensions. Several instruments have been used in the literature to measure the geometrical accuracy of components as shown in Table 6.5.

In recent years, researchers [22, 51, 56, 57] have used laser scanners for measuring the geometrical accuracy of components produced by the ISF process.

Figure 6.7 shows the set-up for measuring geometric accuracy using a blue light 3D laser scanner [22]. The set-up consists of a projection unit with a dual charged coupled device (CCD) camera, an optolink and a computer environment. The projection unit projects fringe patterns on the component to be scanned. The two CCD cameras capture and digitize the fringe patterns reflected from the formed component. The optolink prepares the digitized data for CAD processing using the 3D software.

### 6.4.2 Effects of Parameters on Geometrical Accuracy

Currently, the dimensional accuracy of components produced by the ISF process is a great challenge for researchers and industrial users. The claimed dimensional accuracy for the ISF process is approximately ±3.0 mm for any benchmark shape. Industrial users require geometric accuracy of the order of ±0.2 mm for sheet material components over the whole surface of the part [58].

Since springback is the major issue, it should be targeted for improving the accuracy of components formed by the ISF process. In conventional sheet metal forming operations, springback is observed as the forming punch is removed after completing the forming operation due to the fact that the contact area of the forming punch and

## TABLE 6.5
## Devices Used for Measuring Geometrical Accuracy

| Name of Instrument | Pros | Cons | Reference |
|---|---|---|---|
| Coordinate-measuring machine (CMM) | High accuracy | Processing is slower than that of laser scanner, position of component needs to be specific | [48, 52–54] |
| Argus optical measurement system | Moderate accuracy, high local resolution, can detect critical deformation areas | High cost | [55] |
| ATOS optical measuring device (non-contact 3D scanners using structured blue light) | Moderate accuracy, high local resolution, detects the movement of sensor or part, live tracking of 3D sensor position | High cost | [55] |
| Laser scanner | High accuracy, live tracking of 3D sensor position, non-contact method of measurement, processing is fast | High cost | [22, 51, 56, 57] |

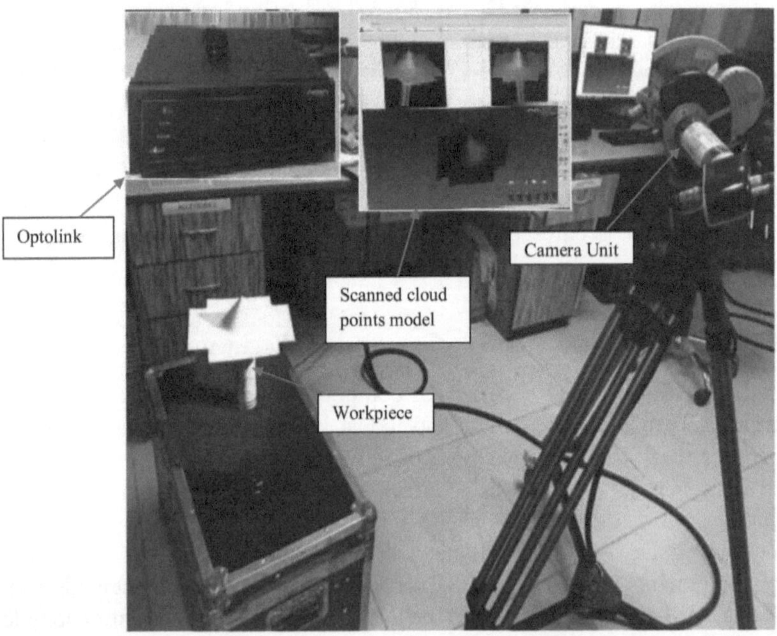


**FIGURE 6.7** Set-up for scanning formed part with a blue light 3D laser scanner [22].

the sheet is very large. Therefore, springback is investigated once the load is removed after completing the conventional forming operation. On the other hand, in ISF, the tool–sheet contact zone is relatively very small and deformation takes place contour by contour, locally. In addition, the incrementally formed components need to be cut or trimmed after the forming operation to achieve the final shape. Basically, the following types of springback are observed in ISF [59]:

a) A continuous local springback (which happens simultaneously with the movement of the forming-tool)
b) A global springback (which happens after removing the forming-force and unclamping the sheet from the fixture)

The ISF process parameters can affect and control the springback greatly and the right selection of process parameters can enhance the geometrical accuracy of components. Table 6.6 gives the concise knowledge to readers about the effect of parameters on the geometrical accuracy of components during the ISF process.

## 6.5 THICKNESS REDUCTION AND THINNING LIMIT

Ambrogio et al. [65] depicted that stretching conditions prevail in the SPIF process and significant thinning of sheets occurs. The maximum allowable thinning of the sheet is an important response of the ISF process. It is mandatory for a process engineer to know the allowable thinning of sheet material without fracture for successful ISF [66].

The final sheet thickness of the component for a particular wall angle in the deformed area varies as per the sine law (Chapter 3). The final thickness of the sheet tends to decrease with the increase in wall angle and become zero with a 90° wall angle, which results in sheet fracture.

If the material of the sheet can be considered incompressible, then the sheet is deformed at the expense of thickness. Hence, thickness reduction, thinning limit and thickness distribution can be considered as important criteria to predict the feasibility of the ISF process [67]. The sheet thickness at which failure occurs is known as the "*thinning limit*" of the sheet material.

### 6.5.1 Measurement of Thickness Reduction or Thinning Limit

The thickness of the formed sheet material of components can be measured by the following two approaches:

a) Manual or destructive methods (using micrometer, digital pointed anvil micrometer, dial gauge indicator, etc.)
b) Non-destructive or advanced methods (using CMM, 3D scanner)

In the manual method of measuring thickness reduction, the formed component is first cut in the middle and then marked along its depth at certain intervals using a depth gauge. Finally, the thickness at the marked points (that is, at different depths) is measured by a digital pointed anvil micrometer or a dial gauge indicator [68], and

**TABLE 6.6**
**Effects of Input Parameters on Geometrical Accuracy in the ISF Process**

| Name of Parameter | Effect on Geometrical Accuracy | Remark | Reference |
|---|---|---|---|
| Tool-path | Helical tool-path produces better parts accuracy as compared to that produced by profile tool-path | The forming-tool remains pressed with the sheet in the helical tool-path and avoids local springback | [60, 61] |
| Tool-path correction method | Algorithms for correcting tool-path using feature-based approach can enhance the accuracy greatly | Wavelet and fast Fourier transforms improve the closed loop trajectory of the toolpath taking springback into account | [51, 62] |
| Tool diameter | Accuracy decreases with the increase in tool diameter | Smaller tool diameter increases concentration at the tool–sheet contact zone and therefore the forming temperature increases which leads to reduced springback | [57, 63] |
| Step size | Accuracy decreases with the increase in step size | Higher step size leads to increment in residual stresses and enhanced springback effect | [52, 56, 57] |
| Feed rate | Accuracy decreases with the increase in feed rate | Fast movement of the forming-tool over the surface of the sheet results in incomplete plastic deformation of the material that increases the springback effect | [57] |
| Spindle speed | Accuracy increases with the increase in tool rotation | Increase in spindle speed results in the increase in friction at the tool–sheet interface that results in increasing the local forming temperature | [57] |
| Forming temperature | HA-ISF processes produce better geometrical accuracy | Increasing temperature of sheet material increases ductility and reduces residual stresses and therefore springback is reduced | [48, 64] |
| Workpiece shape | Geometrical accuracy is greater for conical frustum as compared to that of pyramidal frustum | Pyramidal shape has corner regions that have larger stiffness as compared to oblique walls in conical frustums. More residual stresses are developed in stiffer regions, increasing springback | [57] |
| Forming machinery | Parts made on a CNC milling machine are more accurate than those produced by industrial robots | Accuracy of the component is dependent on the stiffness of the forming set-up. CNC milling machines are stiffer than industrial robots | [24] |

sheet-thinning with increasing depth is calculated by measuring the corresponding sheet thicknesses at marked points.

Some advanced and non-contact type instruments like CMM and 3D scanners can also be used to measure the thickness of the sheet with high accuracy, at any section of the formed component, without sectioning and destroying the components [11].

### 6.5.2 Effects of Parameters on Thickness Reduction and Thinning Limit

The effects of each input parameter on the thickness reduction percentage and the thinning limit of the formed components are required by a process engineer for the successful implementation of the ISF process on an industrial scale. The literature [11, 54, 56,68, 69, 70, 71] reveals that the evaluation of the thinning limit of components is very important in order to ensure safe forming using the ISF process. The minimal value of the thinning limit can be assumed to be optimal for the safe forming of components.

Constant wall angle conical frustums exhibit lower thinning as compared to those with a varying wall angle [71]. Moreover, a helical tool-path produces homogeneous and reduced thickness reduction in the ISF process. Honarpisheh et al. [11] investigated the influences of wall angle and step size on the Ti-6Al-4V sheet at 400–500°C using the electric-assisted ISF process. Results showed that thickness reduction in parts increased with the increase in wall angle and step down size. The thickness distribution of parts was found to be approximately in accordance with the sine law. Attanasio et al. [53] optimized tool-paths and showed that constant scallop height is an efficient tool-path. Moreover, a lower value of step depth and scallop height results in minimum sheet thinning. Blaga and Oleksik [61] investigated the effects of different tool-path strategies on the thickness distribution of cone shape parts from DC04 steel sheets. Tool-paths were profile (constant step depth and step depth with the variation of 90° in every contour) and helical. A decrease of approximately 13.5% was recorded in maximum thickness reduction along depth of the part with the helical tool-path as compared with the other two tool-paths.

Li et al. [67] investigated the impact of tool diameter and tool shape on the thickness distribution of AA7075-O alloy sheets. A grove tests was performed until fracture. Results showed that the thickness formed at certain depth is lower with the decrease in tool diameter which demonstrates earlier fracture in the case of a lower tool diameter.

Kurra et al. [68] investigated thickness distribution on extra deep drawing (EDD) steel sheets to produce varying wall angle pyramidal frustum (VWAPF) of circular, elliptical, parabolic and exponential generatrix. Results showed that EDD sheets could be formed without fracture up to 75% of thickness reduction. Moreover, a higher thickness reduction was noticed at the corners of the pyramid as compared to at the wall section of pyramid. Kurra et al. [69] studied the thickness distribution to produce VWACF of different generatrix (circular, elliptical, parabolic and exponential) on EDD steel sheets. Results showed that the theoretical thickness prediction based on the sine law failed in the bending region. The thinning limit was lower (0.236 mm at 62.5 mm fracture depth) for circular generatrix, whereas it was maximal (0.290 mm at 50 mm fracture depth) for parabolic generatrix. Devarajan et al. [70]

evaluated thickness reduction using a 3D scanner on AA2024-O, AA5083-H111 and AA7075-O sheets. The wall angle was proved to be a limiting factor for the thinning limit. Strain and thinning were higher for steeper wall angles and lead to fracture of the component. Hussain et al. [71] investigated thinning limits and thickness distribution of aluminum sheets experimentally in order to produce VWACF and compared with those predicted by the sine law. Fixed angle conical and pyramidal shapes were also formed to confirm the sine law. Results showed that the thinning limit of varying wall angle cone shapes was higher (approx. 3.87%) than that of fixed-slope cone and pyramid shape. Moreover, the lower possible thinning limit could be attained with a fixed-slope cone shape as compared to those formed at varying wall angle. The thickness distribution was found to follow the sine law well except in bending and fracture regions.

## 6.6 SURFACE ROUGHNESS

The performance of a component can be estimated by its surface roughness. A component may be subjected to corrosion or cracks due to the propagation of irregularities present in the surface. Roughness is the measurement of the vertical variation of the actual surface of the component from the ideal form. The surface roughness of produced components can be represented by many parameters of roughness, but in ISF, most researchers have used average roughness to determine the surface quality of formed parts.

### 6.6.1 Measurement of Surface Roughness

The average roughness ($R_a$) value is the arithmetic mean of the surface profile departure from ideal form. $R_a$ is the average of the peaks and valleys to the mean line for all cut-off lengths. The cut-off length represents the distance traveled by the stylus over the surface in one go (see Figure 6.8). The longer the cut-off length is, the higher the value of the average is. The $R_a$ value of the formed components can be measured by a roughness tester.

Mathematically, the $R_a$ value for a measuring length $L_m$ is given by Equation 6.1, which represents the integral of the height of surface irregularities.

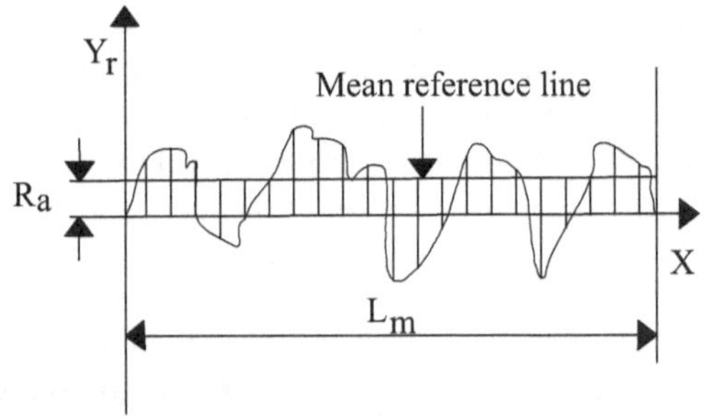


**FIGURE 6.8** Surface roughness profile parameters.

$$R_a = \frac{1}{L_m} \int_0^{L_m} |yr\,(x)|\,\mathrm{dx} \tag{6.1}$$

where *yr* is roughness spread from the reference line and *x* is the profile direction.

### 6.6.2 Effects of Parameters on Surface Roughness

The surface roughness achieved on components by the ISF process is largely dependent on the input parameters of the process. The effects of some input parameters on the surface quality of incrementally formed components are still under debate as the effects of some process parameters are either not very clear or have contradictory results. Most researchers have investigated the internal surface roughness of formed components (i.e. the surface that is formed by direct contact of the forming-tool) whereas some researchers have investigated the quality of outer surface of parts and noticed orange peel effects. Zhang et al. [72] investigated the effects of solid lubricants (solid graphite, K2Ti4O9 and $MoS_2$), and lubricating methods (sprayed, tool-submerged in the lubricant and be-powdered) with and without pulsed anodic oxidation (PAO) in hot incremental forming of magnesium alloy AZ31 sheets on the internal surface roughness. Their results showed that when graphite and $MoS_2$ lubricants were used with a "tool-submerged in the lubricant" type lubricating method, a rough and nicked surface of components was obtained. They also showed that the ceramic coating on the sheets using PAO is also effective and useful for producing a better surface quality. Ajay [22] and Kumar and Gulati [73] investigated the effects of various process factors like sheet thickness, step size, wall angle, tool diameter, tool shape, feed rate, spindle speed and viscosity of lubricating oils (Castrol Alpha SP series) on the surface quality of the inner surface of components formed from AA 2024-O and AA6063-O sheets and optimized the process parameters in terms of surface quality. Results showed that the surface roughness decreases with the increase in tool diameter, spindle speed and viscosity of forming oil, whereas it increases with the increase in sheet thickness, component wall angle and step size as shown in Figures 6.9 and 6.10. A higher tool diameter allows a reduction in waviness on the tool–sheet interface, resulting in better surface quality [73, 74, 75]. The increase in spindle speed enables proper lubrication at the contact zone of the tool and sheet, which reduces scratching on the formed surface and results in decreasing average roughness. It is interesting to note the influence of tool shape and viscosity of forming oil on surface roughness. As the side radius of flat-end tools was increased, the roughness of the surface decreased for all three tool diameters (7.52, 11.60, 15.66 mm) and became minimal for hemispherical-end tools. A similar trend was observed when the viscosity of the forming oil was increased from 68 cSt to 320 cSt. This may be due to the fact that low viscosity forming oils are not able to withstand high forming temperatures and get squeezed out from the tool–sheet contact zone. Whereas highly viscous forming oils do not lose lubricating properties at forming temperatures, do not get squeezed out and provide better surface quality. When the wall angle is increased, a higher lateral area of the forming tool-tip touches the sheet surface, which leads

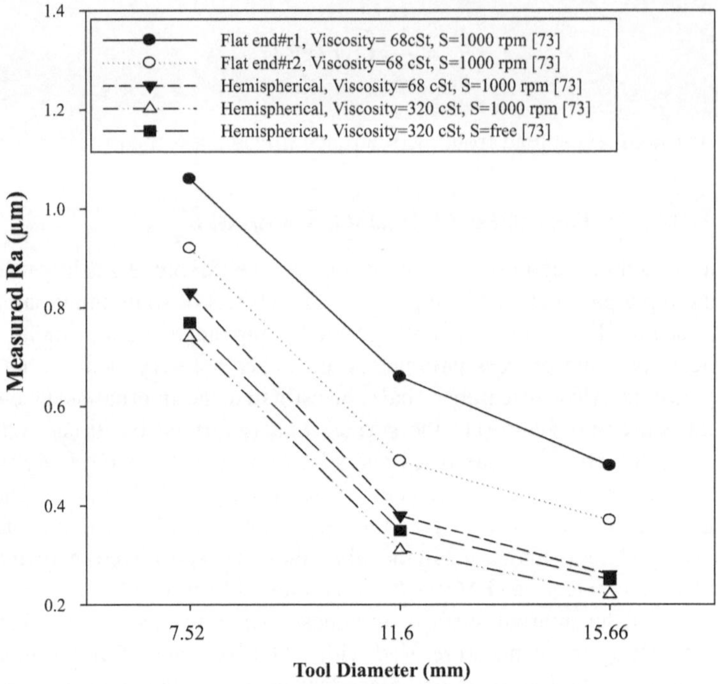


**FIGURE 6.9** Effects of parameters on surface roughness.

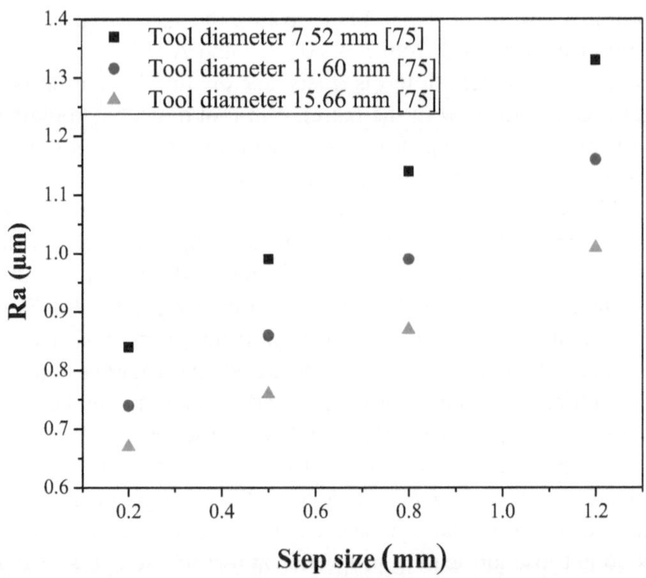


**FIGURE 6.10** Effects of step size and tool diameter on surface roughness.

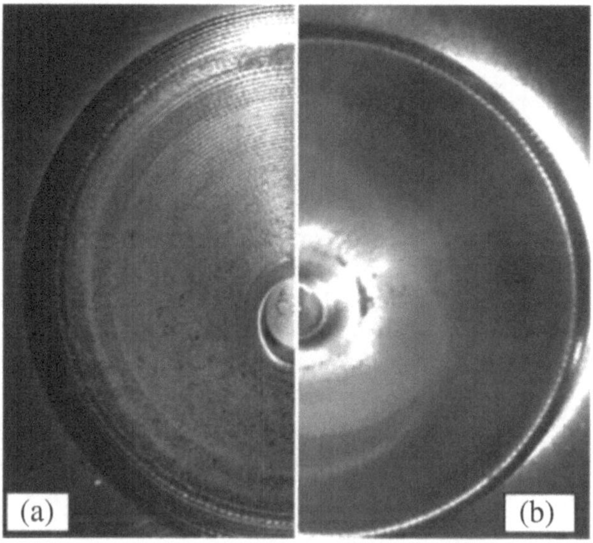


**FIGURE 6.11** Surface quality of ISF component (a) rougher (b) smoother [22].

to an increase in waviness on the surface of the component [74]. A relatively better surface finish was reported with a hemispherical-end tool of diameter 15.66 mm rotating at 1000 RPM deployed with Alpha SP320 forming oil, whereas the surface condition was worst when a flat-end tool of diameter 7.58 mm (Flatend#1 in this case) with a low side radius rotating at 1000 RPM was employed with Alpha SP68 forming oil, as shown in Figure 6.11.

Cavalar et al. [76] studied the impact of tool radius, surface coating of tool and step size on AISI 304 steel sheets and found that the surface roughness decreases with the increase in step size and tool diameter. Coated tools produced a better surface finish than the surface finish produced by uncoated tools. The removal of metal occurred even after using lubricant. The color of the lubricant changed to gray from yellow due to the involvement of particles of metal in it. Babu and Kumar [77] investigated the effects of step depth, spindle speed and feed rate on deep drawing IS513Cr3 steel sheets and found that the surface roughness increased with the increase in spindle speed because the surface became rough due to the increase in temperature. The increase in feed rate and step size resulted in decreasing surface roughness. De Bruyn and Treurnicht [78] investigated the effects of step down and lubricants ($MoS_2$ and graphite) on Ti-6Al-4V sheets. Results showed that the dry run produced a rougher surface of the workpiece as compared to that produced with lubricants. Roughness was found to decrease with the increase in step size. Moreover, the work material tends to adhere to the tool-tip in the case of dry run and reduces the tool life and surface finish.

Oleksik et al. [79] investigated the effects of the tool diameter, the initial average roughness ($R_a$) of the tool (1.25, 2.5 μm) and the friction coefficient (0.08, 0.15) between a sheet and a punch on the average roughness ($R_a$) and maximum roughness ($R_z$) of the inner surface of formed parts. Both $R_a$ and $R_z$ values were found to increase with the increase in tool diameter, initial roughness of tool-tip and the friction coefficient. Moreover, the final inner $R_a$ was more than the initial surface of the blank (0.25 μm), whereas the outer surface of the parts was unaffected by the input parameters.

Hagan and Jeswiet [80] concluded that the surface finish can be viewed as a result of large-scale waviness created by the tool-path and small-scale roughness induced by large surface strains. Powers et al. [81] studied the effects of the rolling direction on the surface morphology of formed components. The results showed that the surface roughness was higher when rolling marks were perpendicular to the forming orientation. Hamilton et al. [82] studied the effects of tool rotation and feed rate on the outer surface roughness (i.e. orange peel effect). They also established a predictive model for orange peel effects of the surface. Either the inner surface or the outer surface has been investigated by researchers at a time during the ISF process. The overall surface quality should be targeted to improve the competitiveness of the components formed by ISF on an industrial scale.

## 6.7 CONCLUDING REMARKS

While this chapter provides a quick overview of some important responses of the ISF process from the available range of publications, it is apparent that ISF is an amenable process that has attracted the attention of leading industries and elite research institutions since the 2000s. The plethora of research articles that have emanated since the initial patent manifests testimony thereof. Special attention has been given to the effects of input parameters on various key responses of the die-less ISF process. The study of the effects of input parameters can become an effective route towards the troubleshooting of various challenges of outputs of this process. Proper understanding and optimization of the forming-force can ensure the safety of forming machinery and forming-tools during the process. A type of threshold in the form of the force gradient has been presented for the readers that can be considered and utilized as a hidden variable for modifying the process parameters by continuous comparison of the crucial value and instant value for safe forming operations. Furthermore, different aspects of formability have been discussed along with their measuring techniques. Forming limit diagrams show that sheet materials can be stretched much more during the ISF process than during conventional forming processes. In addition, the study of the geometrical accuracy of incrementally formed components motivates the researchers and process engineers to achieve the desired tolerance control. Dimensional errors between the target shape and the actual formed shape (due to bending near the clamped region, springback at product walls and the pillow effect at the bottom) can be reduced using optimum process parameters and tool-path. Several advanced methods of measuring the formability and thinning limit of formed components have been described that can directly help the readers to undertake further systematic study to resolve the key challenges of the ISF process. The value added by proper understanding and investigations of process parameters can pave the way for new insights in the different applications of process, unlocking panorama for a wider use of ISF in industrial and academic practice.

## REFERENCES


1. Kumar, Ajay, Vishal Gulati, Parveen Kumar, and Hari Singh. "Forming Force in Incremental Sheet Forming: A Comparative Analysis of the State of the Art." *Journal of the Brazilian Society of Mechanical Sciences and Engineering* 41, no. 6 (2019): 251.
2. Gulati, Vishal, and A. Kumar. "Investigation of Some Process Parameters on Forming Force in Single Point Incremental Forming." *International Research Journal of Engineering and Technology* 4, no. 4 (2018): 784–791.
3. Blaga, Adrian, Octavian Bologa, Valentin Oleksik, and Radu Breaz. "Influence of Tool Path on Main Strains, Thickness Reduction and Forces in Single Point Incremental Forming Process." *Proceedings in Manufacturing Systems* 6, no. 4 (2011): 191–196.
4. Blaga, Adrian, and Valentin Oleksik. "A Study on the Influence of the Forming Strategy on the Main Strains, Thickness Reduction, and Forces in a Single Point Incremental Forming Process." *Advances in Materials Science and Engineering* (2013): 1–10.
5. Thibaud, Sébastien, R. Ben Hmida, Fabrice Richard, and Pierrick Malécot. "A Fully Parametric Toolbox for the Simulation of Single Point Incremental Sheet Forming Process: Numerical Feasibility and Experimental Validation." *Simulation Modelling Practice and Theory* 29 (2012): 32–43.
6. Arfa, H., R. Bahloul, and H. BelHadjSalah. "Finite Element Modelling and Experimental Investigation of Single Point Incremental Forming Process of Aluminum Sheets: Influence of Process Parameters on Punch Force Monitoring and on Mechanical and Geometrical Quality of Parts." *International Journal of Material Forming* 6, no. 4 (2013): 483–510.
7. Liu, Zhaobing, Yanle Li, and Paul Anthony Meehan. "Experimental Investigation of Mechanical Properties, Formability and Force Measurement for AA7075-O Aluminum Alloy Sheets Formed by Incremental Forming." *International Journal of Precision Engineering and Manufacturing* 14, no. 11 (2013): 1891–1899.
8. Bagudanch, I., M.L. Garcia-Romeu, G. Centeno, A. Elías-Zúñiga, and J. Ciurana. "Forming Force and Temperature Effects on Single Point Incremental Forming of Polyvinylchloride." *Journal of Materials Processing Technology* 219 (2015): 221–229.
9. Centeno, Gabriel, Isabel Bagudanch, Andrés Jesús Martínez-Donaire, Maria Luisa Garcia-Romeu, and Carpóforo Vallellano. "Critical Analysis of Necking and Fracture Limit Strains and Forming Forces in Single-Point Incremental Forming." *Materials & Design* 63 (2014): 20–29.
10. Kumar, Ajay, Vishal Gulati, and Parveen Kumar. "Investigation of Process Variables on Forming Forces in Incremental Sheet Forming." *International Journal of Engineering and Technology* 10 (2018): 680–684.
11. Honarpisheh, M., M.J. Abdolhoseini, and S. Amini. "Experimental and Numerical Investigation of the Hot Incremental Forming of Ti-6Al-4V Sheet Using Electrical Current." *The International Journal of Advanced Manufacturing Technology* 83, no. 9–12 (2016): 2027–2037.
12. Kumar, Ajay, and Vishal Gulati. "Experimental Investigations and Optimization of Forming Force in Incremental Sheet Forming." *Sādhanā* 43, no. 10 (2018): 159.
13. Davarpanah, Mohammad Ali, Amin Mirkouei, Xiaoyan Yu, Rajiv Malhotra, and Srikanth Pilla. "Effects of Incremental Depth and Tool Rotation on Failure Modes and Microstructural Properties in Single Point Incremental Forming of Polymers." *Journal of Materials Processing Technology* 222 (2015): 287–300.
14. Lee, B.H., Y.T. Keum, and R.H. Wagoner. "Modeling of the Friction Caused by Lubrication and Surface Roughness in Sheet Metal Forming." *Journal of Materials Processing Technology* 130 (2002): 60–63.
15. Duflou, Joost, Yasemin Tunckol, Alex Szekeres, and Paul Vanherck. "Experimental Study on Force Measurements for Single Point Incremental Forming." *Journal of Materials Processing Technology* 189, no. 1–3 (2007): 65–72.

16. Petek, Aleš, Karl Kuzman, and Janez Kopač. "Deformations and Forces Analysis of Single Point Incremental Sheet Metal Forming." *Archives of Materials science and Engineering* 35, no. 2 (2009): 107–116.
17. Skjødt, Martin, Michael H. Hancock, and Niels Bay. "Creating Helical Tool Paths for Single Point Incremental Forming." *Key Engineering Materials* 344 (2007): 583–590.
18. Wang, Jyhwen, Mahesh Nair, and Ying Zhang. "An Efficient Force Prediction Strategy in Single Point Incremental Sheet Forming." *Procedia Manufacturing* 5 (2016): 761–771.
19. Al-Ghamdi, Khalid A., and G. Hussain. "Forming Forces in Incremental Forming of a Geometry with Corner Feature: Investigation into the Effect of Forming Parameters Using Response Surface Approach." *The International Journal of Advanced Manufacturing Technology* 76, no. 9–12 (2015): 2185–2197.
20. Aerens, Richard, Philip Eyckens, Albert Van Bael, and J.R. Duflou. "Force Prediction for Single Point Incremental Forming Deduced from Experimental and FEM Observations." *The International Journal of Advanced Manufacturing Technology* 46, no. 9–12 (2010): 969–982.
21. Shrivastava, Parnika, and Puneet Tandon. "Investigation of the Effect of Grain Size on Forming Forces in Single Point Incremental Sheet Forming." *Procedia Manufacturing* 2 (2015): 41–45.
22. Ajay. "Optimization of Process Parameters in Incremental Sheet Forming: Experimental and Theoretical Analysis." *PhD Thesis, Guru Jambheshwar University of Science and Technology, Hisar, India,* (2018).
23. Ambrogio, G., L. Filice, and F. Micari. "A Force Measuring Based Strategy for Failure Prevention in Incremental Forming." *Journal of Materials Processing Technology* 177, no. 1–3 (2006): 413–416.
24. Verbert, J. "Computer Aided Process Planning for Rapid Prototyping with Incremental Sheet Forming Techniques." *PhD Thesis*, Katholieke Universiteit Leuven, Leuven (2010).
25. Fratini, L., G. Ambrogio, R. Di Lorenzo, L. Filice, and F. Micari. "Influence of Mechanical Properties of the Sheet Material on Formability in Single Point Incremental Forming." *CIRP Annals* 53, no. 1 (2004): 207–210.
26. Behera, A.K., J. Duflou, and B. Lauwers. "Shape Feature Taxonomy Development for Toolpath Optimisation in Incremental Sheet Forming." *Katholieke Universiteit Leuven: Leuven, Belgium, PhD Thesis* (2013).
27. Bastos, Ricardo N. Pereira, Ricardo J. Alves de Sousa, and Jorge A. Fernandes Ferreira. "Enhancing Time Efficiency on Single Point Incremental Forming Processes." *International Journal of Material Forming* 9, no. 5 (2016): 653–662.
28. Martins, P.A.F., L. Kwiatkowski, V. Franzen, A.E. Tekkaya, and M. Kleiner. "Single Point Incremental Forming of Polymers." *CIRP Annals* 58, no. 1 (2009): 229–232.
29. Jeswiet, J., and D. Young. "Forming Limit Diagrams for Single-Point Incremental Forming of Aluminium Sheet." *Proceedings of the Institution of Mechanical Engineers, Part B: Journal of Engineering Manufacture* 219, no. 4 (2005): 359–364.
30. Lu, B., Y. Fang, D.K. Xu, J. Chen, S. Ai, H. Long, H. Ou, and Jian Cao. "Investigation of Material Deformation Mechanism in Double Side Incremental Sheet Forming." *International Journal of Machine Tools and Manufacture* 93 (2015): 37–48.
31. Filice, L., L. Fratini, and F. Micari. "Analysis of Material Formability in Incremental Forming." *CIRP Annals* 51, no. 1 (2002): 199–202.
32. Shim, Myoung-Sup, and Jong-Jin Park. "The Formability of Aluminum Sheet in Incremental Forming." *Journal of Materials Processing Technology* 113, no. 1–3 (2001): 654–658.
33. Iseki, Hideo. "An Approximate Deformation Analysis and FEM Analysis for the Incremental Bulging of Sheet Metal Using a Spherical Roller." *Journal of Materials Processing Technology* 111, no. 1–3 (2001): 150–154.

34. Li, Yanle, Xiaoxiao Chen, Zhaobing Liu, Jie Sun, Fangyi Li, Jianfeng Li, and Guoqun Zhao. "A Review on the Recent Development of Incremental Sheet-Forming Process." *The International Journal of Advanced Manufacturing Technology* 92, no. 5–8 (2017): 2439–2462.
35. Iseki, H., K. Kato, and S. Sakamoto. "Forming Limit of Flexible and Incremental Sheet Metal Bulging with a Spherical Roller." *Advanced Technical Plastics* 1, no. 1 (1993): 1635–1640.
36. Kim, T.J., and Dong-Yol Yang. "Improvement of Formability for the Incremental Sheet Metal Forming Process." *International Journal of Mechanical Sciences* 42, no. 7 (2000): 1271–1286.
37. Kim, Y.H., and J.J. Park. "Effect of Process Parameters on Formability in Incremental Forming of Sheet Metal." *Journal of Materials Processing Technology* 130 (2002): 42–46.
38. Park, Jong-Jin, and Yung-Ho Kim. "Fundamental Studies on the Incremental Sheet Metal Forming Technique." *Journal of Materials Processing Technology* 140, no. 1–3 (2003): 447–453.
39. Strano, Matteo, Marco Ruggiero, and Luigi Carrino. "Representation of Forming Limits for Negative Incremental Forming of Thin Sheet Metals." *Proceedings of the International Deep Drawing Research Group (IDDRG)* (2004): 198–207.
40. Duflou, J.R., Bart Callebaut, Johan Verbert, and Hans De Baerdemaeker. "Laser Assisted Incremental Forming: Formability and Accuracy Improvement." *CIRP Annals* 56, no. 1 (2007): 273–276.
41. Ji, Y.H., and J.J. Park. "Formability of Magnesium AZ31 Sheet in the Incremental Forming at Warm Temperature." *Journal of Materials Processing Technology* 201, no. 1–3 (2008): 354–358.
42. Fan, Guoqiang, L. Gao, G. Hussain, and Zhaoli Wu. "Electric Hot Incremental Forming: A Novel Technique." *International Journal of Machine Tools and Manufacture* 48, no. 15 (2008): 1688–1692.
43. Hussain, G., L. Gao, and Z.Y. Zhang. "Formability Evaluation of a Pure Titanium Sheet in the Cold Incremental Forming Process." *The International Journal of Advanced Manufacturing Technology* 37, no. 9–10 (2008): 920–926.
44. Ham, M., and J. Jeswiet. "Single Point Incremental Forming and the Forming Criteria for AA3003." *CIRP Annals* 55, no. 1 (2006): 241–244.
45. Kumar, Ajay, Vishal Gulati, Parveen Kumar, Vinay Singh, Brijesh Kumar, and Hari Singh. "Parametric Effects on Formability of AA2024-O Aluminum Alloy Sheets in Single Point Incremental Forming." *Journal of Materials Research and Technology* 8, no. 1 (2019): 1461–1469.
46. Lu, B., Y. Fang, D.K. Xu, J. Chen, H. Ou, N.H. Moser, and Jian Cao. "Mechanism Investigation of Friction-Related Effects in Single Point Incremental Forming Using a Developed Oblique Roller-Ball Tool." *International Journal of Machine Tools and Manufacture* 85 (2014): 14–29.
47. Hirt, G., J. Ames, M. Bambach, and R. Kopp. "Forming Strategies and Process Modelling for CNC Incremental Sheet Forming." *CIRP Annals* 53, no. 1 (2004): 203–206.
48. Duflou, J.R., Bart Callebaut, Johan Verbert, and Hans De Baerdemaeker. "Improved SPIF Performance Through Dynamic Local Heating." *International Journal of Machine Tools and Manufacture* 48, no. 5 (2008): 543–549.
49. Micari, F., G. Ambrogio, and L. Filice. "Shape and Dimensional Accuracy in Single Point Incremental Forming: State of the Art and Future Trends." *Journal of Materials Processing Technology* 191, no. 1–3 (2007): 390–395.
50. Ambrogio, G., I. Costantino, L. De Napoli, L. Filice, L. Fratini, and M. Muzzupappa. "Influence of Some Relevant Process Parameters on the Dimensional Accuracy in Incremental Forming: A Numerical and Experimental Investigation." *Journal of Materials Processing Technology* 153 (2004): 501–507.

51. Fu, Zemin, Jianhua Mo, Fei Han, and Pan Gong. "Tool Path Correction Algorithm for Single-Point Incremental Forming of Sheet Metal." *The International Journal of Advanced Manufacturing Technology* 64, no. 9–12 (2013): 1239–1248.
52. Fiorentino, Antonio, Elisabetta Ceretti, Aldo Attanasio, Luca Mazzoni, and Claudio Giardini. "Analysis of Forces, Accuracy and Formability in Positive Die Sheet Incremental Forming." *International Journal of Material Forming* 2, no. 1 (2009): 805.
53. Attanasio, Aldo, Elisabetta Ceretti, and Claudio Giardini. "Optimization of Tool Path in Two Points Incremental Forming." *Journal of Materials Processing Technology* 177, no. 1–3 (2006): 409–412.
54. Cerro, Itsasne, Elixabete Maidagan, José Arana, A. Rivero, and P.P. Rodriguez. "Theoretical and Experimental Analysis of the Dieless Incremental Sheet Forming Process." *Journal of Materials Processing Technology* 177, no. 1–3 (2006): 404–408.
55. Franzen, V., L. Kwiatkowski, G. Sebastiani, R. Shankar, A.E. Tekkaya, and M. Kleiner. "Dyna-Die: Towards Full Kinematic Incremental Forming." *International Journal of Material Forming* 1, no. 1 (2008): 1163–1166.
56. Malhotra, Rajiv, Liang Xue, Ted Belytschko, and Jian Cao. "Mechanics of Fracture in Single Point Incremental Forming." *Journal of Materials Processing Technology* 212, no. 7 (2012): 1573–1590.
57. Radu, M.C., and I. Cristea. "Processing Metal Sheets by SPIF and Analysis of Parts Quality." *Materials and Manufacturing Processes* 28, no. 3 (2013): 287–293.
58. Allwood, J.M., G.P.F. King, and Joost Duflou. "A Structured Search for Applications of the Incremental Sheet-Forming Process by Product Segmentation." *Proceedings of the Institution of Mechanical Engineers, Part B: Journal of Engineering Manufacture* 219, no. 2 (2005): 239–244.
59. Gatea, Shakir, Hengan Ou, and Graham McCartney. "Review on the Influence of Process Parameters in Incremental Sheet Forming." *The International Journal of Advanced Manufacturing Technology* 87, no. 1–4 (2016): 479–499.
60. Malhotra, Rajiv, N.V. Reddy, and Jian Cao. "Automatic 3D Spiral Toolpath Generation for Single Point Incremental Forming." *Journal of Manufacturing Science and Engineering* 132, no. 6 (2010): 061003.
61. Blaga, Adrian, and Valentin Oleksik. "A Study on the Influence of the Forming Strategy on the Main Strains, Thickness Reduction, and Forces in a Single Point Incremental Forming Process." *Advances in Materials Science and Engineering* (2013): 1–10.
62. Lu, B., J. Chen, H. Ou, and Jian Cao. "Feature-Based Tool Path Generation Approach for Incremental Sheet Forming Process." *Journal of Materials Processing Technology* 213, no. 7 (2013): 1221–1233.
63. Oleksik, Valentin, Adrian Pascu, CristianDeac, Radu Fleaca, Mihai Roman, and Octavian Bologa. "The Influence of Geometrical Parameters on the Incremental Forming Process for Knee Implants Analyzed by Numerical Simulation." *AIP Conference Proceedings* 1252, no. 1 (2010): 1208–1215.
64. Mohammadi, Amirahmad, Hans Vanhove, Albert Van Bael, and Joost R. Duflou. "Towards Accuracy Improvement in Single Point Incremental Forming of Shallow Parts Formed Under Laser Assisted Conditions." *International Journal of Material Forming* 9, no. 3 (2016): 339–351.
65. Ambrogio, G., L. Filice, L. Fratini, and F. Micari. "Some Relevant Correlations Between Process Parameters and Process Performance in Incremental Forming of Metal Sheets." In: *Proceedings of the 6th Esaform Conference on Material forming*, Salerno, Italy, April 28–30, 2003, pp. 175–178, 2003.
66. Young, D., and J. Jeswiet. "Wall Thickness Variations in Single-Point Incremental Forming." *Proceedings of the Institution of Mechanical Engineers, Part B: Journal of Engineering Manufacture* 218, no. 11 (2004): 1453–1459.

67. Li, Yanle, Zhaobing Liu, W.J.T. Daniel, and P.A. Meehan. "Simulation and Experimental Observations of Effect of Different Contact Interfaces on the Incremental Sheet Forming Process." *Materials and Manufacturing Processes* 29, no. 2 (2014): 121–128.
68. Suresh, Kurra, SreeDivya Bagade, and Srinivasa Prakash Regalla. "Deformation Behavior of Extra Deep Drawing Steel in Single-Point Incremental Forming." *Materials and Manufacturing Processes* 30, no. 10 (2015): 1202–1209.
69. Kurra, Suresh, and Srinivasa Prakash Regalla. "Experimental and Numerical Studies on Formability of Extra-Deep Drawing Steel in Incremental Sheet Metal Forming." *Journal of Materials Research and Technology* 3, no. 2 (2014): 158–171.
70. Devarajan, Nagarajan, Giribaskar Sivaswamy, Rahul Bhattacharya, David P. Heck, and Muhammad Amir Siddiq. "Complex Incremental Sheet Forming Using Back Die Support on Aluminium 2024, 5083 and 7075 Alloys." *Procedia Engineering* 81 (2014): 2298–2304.
71. Hussain, G., and L. Gao. "A Novel Method to Test the Thinning Limits of Sheet Metals in Negative Incremental Forming." *International Journal of Machine Tools and Manufacture* 47, no. 3–4 (2007): 419–435.
72. Zhang, Qinglai, Fugui Xiao, Hailing Guo, Changsheng Li, Lin Gao, Xingwu Guo, Weidong Han, and A.B. Bondarev. "Warm Negative Incremental Forming of Magnesium Alloy AZ31 Sheet: New Lubricating Method." *Journal of Materials Processing Technology* 210, no. 2 (2010): 323–329.
73. Kumar, Ajay, and Vishal Gulati. "Experimental Investigation and Optimization of Surface Roughness in Negative Incremental Forming." *Measurement* 131 (2019): 419–430.
74. Kumar, Ajay, Vishal Gulati, and Parveen Kumar. "Effects of Process Parameters on Surface Roughness in Incremental Sheet Forming." *Materials Today: Proceedings* 5, no. 14 (2018): 28026–28032.
75. Kumar, Ajay, Vishal Gulati, and Parveen Kumar. "Investigation of Surface Roughness in Incremental Sheet Forming." *Procedia Computer Science* 133 (2018): 1014–1020.
76. Cavaler, L.C.C., L. Schaeffer, A.S. Rocha, and F. Peruch. "Surface Roughness in the Incremental Forming of AISI 304L Stainless Steel Sheets." *Journal of Mechanical Engineering Physics* 1, no. 2 (2010): 87–98.
77. Babu, S. Chezhian, and V.S. Senthil Kumar. "Effect of Process Variables During Incremental Forming of Deep Drawing Steel Sheets." *European Journal of Scientific Research* 80 (2012): 50–56.
78. de Bruyn, Ruan, and Nico Treurnicht. "An Investigation into Lubrication Strategies for the Incremental Sheet Forming of Ti-6Al-4V." In: Proceedings of the 42nd international conference on computers and industrial engineering, Cape Town, South Africa, 15–18 July 2012, pp. 193–200.
79. Oleksik, V., A. Pascu, C. Deac, R. Fleacă, O. Bologa, and G. Racz. "Experimental Study on the Surface Quality of the Medical Implants Obtained by Single Point Incremental Forming." *International Journal of Material Forming* 3, no. 1 (2010): 935–938.
80. Hagan, E., and J. Jeswiet. "Analysis of Surface Roughness for Parts Formed by Computer Numerical Controlled Incremental Forming." *Proceedings of the Institution of Mechanical Engineers, Part B: Journal of Engineering Manufacture* 218, no. 10 (2004): 1307–1312.
81. Powers, B.M., M. Ham, and M.G. Wilkinson. "Small Data Set Analysis in Surface Metrology: An Investigation Using a Single Point Incremental Forming Case Study." *Scanning* 32, no. 4 (2010): 199–211.
82. Hamilton, Kelvin Allan Samuel. "Friction and External Surface Roughness in Single Point Incremental Forming: A Study of Surface Friction, Contact Area and the 'Orange Peel' Effect." *PhD dissertation, Queen's University, Ontario, Canada* (2010).

# 7 Applications and Challenges of Incremental Sheet Forming Processes

## 7.1 INTRODUCTION

This chapter focuses on viable applications of the ISF process including medical, automotive, dies and molds, aerospace, architecture, prototyping and transportation applications. As ISF has been developed over the years, various applications have been documented in the literature. However, the direct uses of ISF in the form of functional parts are in small numbers, but the variety of different complex shapes has been increasing, showing that this method of forming is gradually developing to prove its significance in the market [1, 2]. The direct forming of functional parts has been accomplished by researchers in small numbers, basically ranging from single tailored parts to a series of 10–15 of parts. Some published documents have shown that the breakeven point of the ISF process is a few hundred parts. Cooper et al. [3] concluded that the breakeven point of the ISF process can be only a hundred parts as it avoids the manufacturing of dedicated dies that can save energy per part under consideration. Furthermore, the accuracy of parts formed by the ISF process has been increased up to sub-millimeter tolerances, enabling demonstrations of functional parts in growing numbers of application areas as discussed in the following sections. The various challenges and issues that limit the applications of this process have also been discussed in this chapter. The proper understanding and knowledge of challenges can help researchers and industrial personnel to take the opportunity to enhance the applicability and sustainability of the process on an industrial scale [4, 5, 6, 7].

## 7.2 APPLICATIONS OF THE ISF PROCESS

Applications of the ISF process are vast and diverse and are not limited to industrial products. Typical applications in the medical, architectural, die-making and prototyping areas are discussed here.

### 7.2.1 Bio-Medical and Medical Applications

ISF has become a well-positioned process for producing patient-orientated medical components and devices over the years. The ISF process is suitable for medical

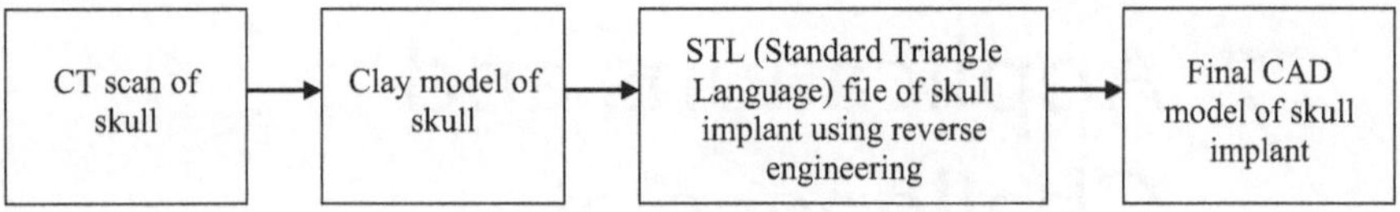


**FIGURE 7.1** Procedure for producing a CAD model of skull implants.

implants due to the need for customization to the shape of the human body. The following are specific applications for implants [8]:

- Ankle support
- Cranial plate
- Facial implant
- Knee prosthesis
- Palate prosthesis
- Backseat prosthesis

Different materials like titanium grade 1 [9] and grade 2 [10], polycaprolactone [11] and nano-polymer composites [12] have been experimented with. Ambrogio et al. [13] produced a patient-specific foot orthosis using the ISF process. Figure 7.1 shows the procedure for producing a CAD model of skull implants as reported by Duflou et al. [14].

This procedure of producing skull implants using the ISF process has been recognized by several research groups [1]. The effectiveness of the ISF process in the medical field has been recently demonstrated by Vanhove et al. [15] by producing titanium clavicular implants. McAllister and Jeswiet [16] reported that a multi-step approval process is required for making ISF parts ready for medical use.

### 7.2.2 Architectural Applications

In architectural applications, dimensional accuracy is not stringent (mm range) but the surface quality is important. ISF has capabilities to form large-size double curved surfaces, and this is an addition to the traditional architectural lining methods like rubber forming. Bailly et al. [17] produced a self-bearing structure, and Jackson et al. [18] explored the potential use of sandwich panels by forming 3D shells using the ISF process. Sandwich panels also find their applications as automotive components like dash panels and inner wheel housing.

In architectural applications, the requirement of higher wall angles for a freeform shape is one of the key issues that occurs during the ISF process. Mesh morphing methods can be a solution to form freeform shapes while overcoming the above challenges [18].

### 7.2.3 Die-Making Applications

Various thermoforming methods like vacuum forming need specific die for producing sheet material components. In vacuum forming, plastic parts can be formed by first heating the plastic sheet and then forming the plastic sheet using a die. ISF can be used for making die of sheet metal instead of making a solid die. The die of sheet metal, may need a support that can be a backing plate with porous granular structure, to ensure proper rigidity of the die surface under vacuum conditions [19]. Padrao [20] recently formed a die for a shower basin using the ISF process that can be used as a mold for thermoforming.

### 7.2.4 Prototyping Applications

ISF has a wide range of applications in the automotive, aerospace and transportation sectors for making prototypes, especially where customization is very much required [21–23]. While forming components for these application areas, the working range of size of milling machines commonly available is not sufficient due to the larger size required for parts. This issue is resolved by using industrial robots but at the expense of the dimensional accuracy of the final component due to the lower stiffness of the robots.

Some attempts have been made to show the capabilities of this process to form a solar cooker [24], an engine heat shield [25], a car fender [26], scaled airfoils [27], taillight bracket of a car [28] and the cover of a motorcycle exhaust pipe [29]. Table 7.1 presents a summary of applications of the ISF process in different areas.

**TABLE 7.1**
**Summary of Applications of the ISF Process**

| Application Area | Material | Specific Application | Reference(s) |
|---|---|---|---|
| **Medical** | Polycaprolactone (PCL), Ti grade 2 | Palate prosthesis | [11] |
| | DDQ steel | Ankle support | [13] |
| | AA3003-O, Ti grade 2, PCL | Cranial plate | [10, 14, 30, 31] |
| | Ti grade 2, Pure Ti | Knee prosthesis | [32–34] |
| | AA 3103 | Backseat orthosis | [35] |
| | Ti grade 2 | Facial implant | [36] |
| **Architectural** | – | Self-bearing structure | [17] |
| | Multiple | Sandwich panel | [18] |
| **Dies and mold** | DC01,AA 3103 | Mold | [19] |
| | Stainless steel | Mold | [20] |
| | Al alloy | Mold | [37] |
| **Automotive** | AA 3003-O | Engine heat shield | [25] |
| | AA 3103, DC04 | Car fender | [26, 38] |
| | DC04 | Car tail light | [28] |
| | AA 1100 | Cover of a motorcycle exhaust pipe | [29] |
| | Al alloy | Car body | [39] |
| **Aerospace** | AA 5754 | Airfoils | [40] |

## 7.3 CHALLENGES AND ISSUES OF ISF PROCESS

Although a substantial amount of research has been performed on the ISF process in the past two decades to develop a fundamental understanding and new tooling solutions, the process still needs various insights to be implemented in high-value manufacturing industries due to various technical challenges and issues that are still unsolved.

The optimization of various responses of the conventional ISF process like forming force, surface roughness, productivity, formability, geometric accuracy and thinning limits of the material has been a major focus of researchers in the early stages of development of the process. The optimization of various input parameters and several HA-ISF processes have been another thrust area over the years to enhance the capabilities and application areas of the ISF process. Several issues have been targeted to bridge the gap between process limitations and possibilities of adaptation of the ISF process for industrial rollout. Various tool-path optimization techniques have enabled the process to produce complex shapes with close tolerances, and this has increased the competitiveness of the ISF process. Similarly, HA-ISF techniques have allowed the extension of this success to more robustness and accuracy of the process. Nevertheless, some obstacles remain as challenging bottlenecks to implementing the process on an industrial scale. These are enumerated below:

- Forming time is much greater than with conventional sheet forming processes, and this affects the applicability of the process adversely.
- The accuracy of the final product is inferior to that required due to a larger amount of springback and unwanted bending.
- The formation of vertical walls (close to a 90-degree wall angle) is quite difficult and leads to failure.
- The surface finish of the formed components is poor as compared to other conventional sheet forming processes.
- Material thinning is excessive at higher wall angles and leads to unwanted fracture of the material.
- A lack of analytical models and methods to predict the forming force, surface finish, accuracy, etc., leads to experimentation and simulation, resulting in an increase in costs and set-up time.
- The selection of optimal forming machinery is still an unsolved area as CNC milling machines are not designed for forming components.
- The interaction of process parameters and their impact on ISF are still major challenges for researchers.

The formability of hard-to-form materials can be enhanced using HA-ISF processes. Hence, these methods of forming should be optimized in terms of input parameters and tool-path strategies. Moreover, the behavior of the material and deformation mechanism in hot forming is still a challenge to be met. The dimensional accuracy achieved during the ISF process is also a barrier to implementing the process in the mainstream of industries. On-line tool-path correction techniques have not been implemented well to control unwanted springback and bending during incremental forming. The dimensional inaccuracy of the ISF process is a major limitation and challenge for researchers for the further development and commercialization of the process. The surface finish of formed components is another weak point of the process and a technical problem for researchers. The optimization of tool-paths and forming strategies to enhance the surface finish of components is another research problem [41–43].

The product quality can also be explained in terms of the final thickness of the component at various points after completing the ISF operation. The excessive thinning at higher forming angles has limited the geometrical aspects of this process. Although process parameters have been investigated for conventional ISF and HA-ISF methods, the lack of modeling and analytical methods to predict thickness, forming-force, surface finish, accuracy, etc., leads to experimental and simulation tests, resulting in an increase in forming time and use of energy. Researchers are using CNC milling machines and industrial robots as forming machinery in the ISF process. These machines are not designed as machine tools for the ISF process and, therefore, are neither economical nor ideal for the ISF process. Inappropriate forming machinery worsens the energy efficiency drastically.

## 7.4 CONCLUDING REMARKS

Despite substantial growth in the field of ISF, a great deal of scope for further development still remains. ISF is a green choice in the field of rapid prototyping and small batch size production. It saves as much as 10% of material compared to other conventional forming processes as it directly avoids the use of forming dies. The applications of the ISF process are very wide including the aerospace, automotive, architecture, medical and prototyping fields. It is important to increase the interest of leading industries to target more challenging areas to drive a trend towards the complete commercial success of this technique. The current challenges and issues derived from consistent and continuous investigations should be considered to transform this process from a potential tool to a powerful weapon in the armory of process engineers to a commercial technology.

## REFERENCES


1. Duflou, Joost R., Anne-Marie Habraken, Jian Cao, Rajiv Malhotra, Markus Bambach, Dave Adams, Hans Vanhove, Amirahmad Mohammadi, and Jack Jeswiet. "Single Point Incremental Forming: State-of-the-Art and Prospects." *International Journal of Material Forming* 11, no. 6 (2018): 743–773.
2. Kumar, Ajay, Vishal Gulati, Parveen Kumar, and Hari Singh. "Forming Force in Incremental Sheet Forming: A Comparative Analysis of the State of the Art." *Journal of the Brazilian Society of Mechanical Sciences and Engineering* 41, no. 6 (2019): 251.
3. Cooper, Daniel R., Kathleen E. Rossie, and Timothy G. Gutowski. "The Energy Requirements and Environmental Impacts of Sheet Metal Forming: An Analysis of Five Forming Processes." *Journal of Materials Processing Technology* 244 (2017): 116–135.

4. Allwood, J.M., G.P.F. King, and Joost Duflou. "A Structured Search for Applications of the Incremental Sheet-Forming Process by Product Segmentation." *Proceedings of the Institution of Mechanical Engineers, Part B: Journal of Engineering Manufacture* 219, no. 2 (2005): 239–244.
5. Gulati, Vishal, A. Kumar "Investigation of Some Process Parameters on Forming Force in Single Point Incremental Forming." *International Research Journal of Engineering and Technology* 4, no. 4 (2018): 784–791.
6. Kumar, Ajay, Vishal Gulati, and Parveen Kumar. "Investigation of Process Variables on Forming Forces in Incremental Sheet Forming." *International Journal of Engineering and Technology* 10 (2018): 680–684.
7. Kumar, Ajay, Vishal Gulati, and Parveen Kumar. "Experimental Investigation of Forming Forces in Single Point Incremental Forming." In: Shanker Kripa, Shankar Ravi, & Sindhwani Rahul (eds.), *Advances in Industrial and Production Engineering*, pp. 423–430. Springer, 2019.
8. Behera, Amar Kumar, Ricardo Alves de Sousa, Giuseppe Ingarao, and Valentin Oleksik. "Single Point Incremental Forming: An Assessment of the Progress and Technology Trends from 2005 to 2015." *Journal of Manufacturing Processes* 27 (2017): 37–62.
9. Behera, Amar Kumar, Bin Lu, and Hengan Ou. "Characterization of Shape and Dimensional Accuracy of Incrementally Formed Titanium Sheet Parts with Intermediate Curvatures Between Two Feature Types." *The International Journal of Advanced Manufacturing Technology* 83, no. 5–8 (2016): 1099–1111.
10. Duflou, Joost R., Amar Kumar Behera, Hans Vanhove, and Liciane S. Bertol. "Manufacture of Accurate Titanium Cranio-Facial Implants with High Forming Angle Using Single Point Incremental Forming." *Key Engineering Materials* 549 (2013): 223–230.
11. Fiorentino, A., G.P. Marenda, R. Marzi, E. Ceretti, D.T. Kemmoku, and J.V. Lopes da Silva. "Rapid Prototyping Techniques for Individualized Medical Prosthesis Manufacturing." In: *Proceedings of the 5th International Conference on Advanced Research and Rapid Prototyping*, Leiria, Portugal, 28 September–1 October, 2011, pp. 589–594.
12. Garcia-Romeu, M.L., I. Bagudanch, L.M. Lozano-Sánchez, O. Martínez-Romero, A. Elías-Zúñiga, L.E. Elizalde. "On the Manufacture of Biomedical Devices Based on Nano-Polymer Composites by Single Point Incremental Forming." In: *2nd International Conference on Design and Processes for Medical Devices (PROMED 2014)*, Monterrey, Mexico, edited by C.A. Rodriguez, pp. 49–52, 2014.
13. Ambrogio, G., L. De Napoli, L. Filice, F. Gagliardi, and M. Muzzupappa. "Application of Incremental Forming Process for High Customised Medical Product Manufacturing." *Journal of Materials Processing Technology* 162 (2005): 156–162.
14. Duflou, J.R., B. Lauwers, J. Verbert, F. Gelaude, and Y. Tunckol. "Medical Application of Single Point Incremental Forming: Cranial Plate." In: *Virtual Modelling and Rapid Manufacturing: Advanced Research in Virtual and Rapid Prototyping Proc. 2nd Int. Conf. on Advanced Research in Virtual and Rapid Prototyping*, 28 Sep–1 Oct 2005, Leiria, Portugal, p. 161. CRC Press, 2005.
15. Vanhove, Hans, Yannick Carette, Sanne Vancleef, and Joost R. Duflou. "Production of Thin Shell Clavicle Implants Through Single Point Incremental Forming." *Procedia Engineering* 183 (2017): 174–179.
16. McAllister, P., and J. Jeswiet. "Medical Device Regulation for Manufacturers." *Proceedings of the Institution of Mechanical Engineers, Part H: Journal of Engineering in Medicine* 217, no. 6 (2003): 459–467.
17. Bailly, David, Markus Bambach, Gerhard Hirt, Thorsten Pofahl, Ralf Herkrath, and Martin Trautz. "Analysis into Differences Between the Buckling in Single-Point and Two-Point Incremental Sheet Forming of Components for Self-Supporting Sheet Metal Structures." *Key Engineering Materials* 622 (2014): 367–374.

18. Jackson, K.P., J.M. Allwood, and M. Landert. "Incremental Forming of Sandwich Panels." *Journal of Materials Processing Technology* 204, no. 1–3 (2008): 290–303.
19. Appermont, Raf, Bart Van Mieghem, Albert Van Bael, JanBens, Jan Ivens, Hans Vanhove, Amar Kumar Behera, and Joost Duflou. "Sheet-Metal Based Molds for Low-Pressure Processing of Thermoplastics." In: *Proceedings of the 5th Bi-annual PMI Conference*, vol. 5, pp. 383–388. University College Ghent, Ghent, 2012.
20. Padrao, J.L. "Single Point Incremental Forming." *MSc Thesis*. Instituto Superior Tecnico Universidade Tecnica de Lisboa (2009).
21. Ajay. "Optimization of Process Parameters in Incremental Sheet Forming: Experimental and Theoretical Analysis." *PhD Thesis, Guru Jambheshwar University of Science and Technology, Hisar, India* (2018).
22. Kumar, Ajay, and Vishal Gulati. "Experimental Investigations and Optimization of Forming Force in Incremental Sheet Forming." *Sādhanā* 43, no. 10 (2018): 159.
23. Kumar, Ajay, Vishal Gulati, Parveen Kumar, Vinay Singh, Brijesh Kumar, and Hari Singh. "Parametric Effects on Formability of AA2024-O Aluminum Alloy Sheets in Single Point Incremental Forming." *Journal of Materials Research and Technology* 8, no. 1 (2019): 1461–1469.
24. Jeswiet, J., Joost R. Duflou, Alexander Szekeres, and Pierre Lefebvre. "Custom Manufacture of a Solar Cooker–A Case Study." *Advanced Materials Research* 6 (2005): 487–492.
25. Young, D., and J. Jeswiet. "Wall Thickness Variations in Single-Point Incremental Forming." *Proceedings of the Institution of Mechanical Engineers, Part B: Journal of Engineering Manufacture* 218, no. 11 (2004): 1453–1459.
26. Verbert, J. "Computer Aided Process Planning for Rapid Prototyping with Incremental Sheet Forming Techniques." *PhD Thesis*, Katholieke Universiteit Leuven, Leuven (2010).
27. Behera, Amar Kumar, Bert Lauwers, and Joost R. Duflou. "Tool Path Generation for Single Point Incremental Forming Using Intelligent Sequencing and Multi-Step Mesh Morphing Techniques." *Key Engineering Materials* 554 (2013): 1408–1418.
28. Junchao, Li, Shen Junjian, and Wang Bin. "A Multipass Incremental Sheet Forming Strategy of a Car Taillight Bracket." *The International Journal of Advanced Manufacturing Technology* 69, no. 9–12 (2013): 2229–2236.
29. Benítez Lozano, Adrián José, Gabriel Jaime Páramo Bermúdez, and Frank Alexander Bustamante Correa. "Comparative Analysis Between the Spif and Dpif Variants for Die-Less Forming Process for an Automotive Workpiece." *INGE CUC* 11, no. 2 (2015): 68–73.
30. Fiorentino, A., R. Marzi, and E. Ceretti. "Preliminary Results on Ti Incremental Sheet Forming (ISF) of Biomedical Devices: Biocompatibility, Surface Finishing and Treatment." *International Journal of Mechatronics and Manufacturing Systems* 5, no. 1 (2012): 36–45.
31. Bagudanch, Isabel, Luis Marcelo Lozano-Sánchez, Laura Puigpinós, Marc Sabater, Luis Ernesto Elizalde, Alex Elías-Zúñiga, and Maria Luisa Garcia-Romeu. "Manufacturing of Polymeric Biocompatible Cranial Geometry by Single Point Incremental Forming." *Procedia Engineering* 132 (2015): 267–273.
32. Eksteen, P.D., and A.F. Van der Merwe. "Incremental Sheet Forming (ISF) in the Manufacturing of Titanium Based Plate Implants in the Bio-Medical Sector." *CIE42 Proceedings* 131 (2012): 16–18.
33. Eksteen, P.D. "Development of Incrementally Formed Patient-Specific Titanium Knee Prosthesis." *MSc Thesis*, Stellenbosch University (2013).
34. Oleksik, Valentin, Adrian Pascu, Cristian Deac, Radu Fleaca, Mihai Roman, and Octavian Bologa. "The Influence of Geometrical Parameters on the Incremental Forming Process for Knee Implants Analyzed by Numerical Simulation." *AIP Conference Proceedings* 1252, no. 1 (2010): 1208–1215.

35. Behera, A.K., J. Duflou, and B. Lauwers. “Shape Feature Taxonomy Development for Tool Path Optimisation in Incremental Sheet Forming.” *PhD Thesis*, Katholieke Universiteit Leuven, Leuven, Belgium (2013).
36. Araújo, Rogério, Pedro Teixeira, Luciana Montanari, Ana Reis, Maria B. Silva, and Paulo A.F. Martins. “Single Point Incremental Forming of a Facial Implant.” *Prosthetics and Orthotics International* 38, no. 5 (2014): 369–378.
37. Allwood, J.M., Alan N. Bramley, T.W. Ridgman, and A.R. Mileham. “A Novel Method for the Rapid Production of Inexpensive Dies and Moulds with Surfaces Made by Incremental Sheet Forming.” *Proceedings of the Institution of Mechanical Engineers, Part B: Journal of Engineering Manufacture* 220, no. 2 (2006): 323–327.
38. Bambach, Markus, B. Taleb Araghi, and Gerhard Hirt. “Strategies to Improve the Geometric Accuracy in Asymmetric Single Point Incremental Forming.” *Production Engineering* 3, no. 2 (2009): 145–156.
39. Governale, A., A. Lo Franco, A. Panzeca, Livan Fratini, and Fabrizio Micari. “Incremental Forming Process for the Accomplishment of Automotive Details.” *Key Engineering Materials* 344 (2007): 559–566.
40. Behera, Amar Kumar, Bert Lauwers, and Joost R. Duflou. “Tool Path Generation Framework for Accurate Manufacture of Complex 3D Sheet Metal Parts Using Single Point Incremental Forming.” *Computers in Industry* 65, no. 4 (2014): 563–584.
41. Kumar, Ajay, and Vishal Gulati. “Experimental Investigation and Optimization of Surface Roughness in Negative Incremental Forming.” *Measurement* 131 (2019): 419–430.
42. Kumar, Ajay, Vishal Gulati, and Parveen Kumar. “Effects of Process Parameters on Surface Roughness in Incremental Sheet Forming.” *Materials Today: Proceedings* 5, no. 14 (2018): 28026–28032.
43. Kumar, Ajay, Vishal Gulati, and Parveen Kumar. “Investigation of Surface Roughness in Incremental Sheet Forming.” *Procedia Computer Science* 133 (2018): 1014–1020.

# 8 Conclusions and Future Scope

## 8.1 INTRODUCTION

ISF processes are game-changing manufacturing methods that have the capability to remarkably expand the frontier of sheet forming techniques and the use of sheet material parts. Every sector that requires applications of sheet material components is a candidate for these die-less and flexible forming techniques. The applications of ISF include, but are not limited to, the medical, aerospace, automobile, transportation, architecture and rapid prototyping sectors. Each variant of ISF has its own benefits and limitations. However, some variants of ISF (like DSIF, HA-ISF) are in their infancy and have significant scope and require efforts to become mature and effective substitutes as industrially acceptable forming processes. The forming hardware and input parameters of the ISF process are of great importance and have been explained in the book. The responses of the ISF process determine the suitability and applicability of the process in various sectors and the same has been discussed in detail. A discussion has been given at the end of every relevant chapter. However, some points that need to be specially highlighted are included in the following section.

## 8.2 SUMMARY

ISF techniques can become an effective path to overcome various limitations in the field of sheet forming. ISF is a magnificent technique to fabricate intricate sheet material components for various applications. ISF is a flexible forming technique with industry 4.0 readiness and is awaiting metamorphosis from the labs to manufacturing environments. The following are the key advantages of using ISF techniques as compared to conventional sheet forming techniques: Negligible cost of set-up as compared to the traditional sheet forming set-ups, economical and flexible process, non-existence of typical dies, higher formability, lower forming-force requirement and so on.

The key inferences from the discussions in this book are listed here.

- Various advancements and corresponding challenges in ISF technologies have led to a conscientious pursuit among researchers and various large automotive and commercial companies to look for solutions.
- Conventional ISF techniques such as SPIF, TPIF and DSIF have their own advantages, limitations and specific areas of applications.
- HA-ISF techniques have been devised to provide solutions to the issues encountered in conventional ISF techniques.

- All variants of ISF techniques have been illustrated with their key advantages and limitations.
- The development of machinery and hardware needed to accomplish the ISF process leads to the fabrication of various realistic accurate objects of various materials.
- The forming set-up of ISF consists of forming machinery, forming-tools, lubricants and CAD/CAM software.
- The various forming machines such as CNC milling machines, industrial robots and purpose-built machines have been described and compared, taking into account stiffness, load capability, working area, feed rate and flexibility.
- The effects of the shape of forming-tools such as hemispherical-end, flat-end, VRB and ORB on geometrical accuracy, forming-force, formability and surface roughness of components have inspired researchers to set guidelines to make the ISF process commercial.
- The use of a backing plate in ISF fixtures leads to improved geometrical accuracy of the components as it avoids unnecessary bending near the clamping area.
- Purpose-built machines such as the *Amino ISF machine*, *Cambridge ISF machine* and *SPIF-A machine* are equipped with various online measuring instruments that increase the effectiveness and competitiveness of the ISF process.
- Purpose-built machines have the capacity to access the workpiece from both sides, change the forming-tool with ease, apply high load to deform hard and thick sheets and measure forming-forces, strains and formability online.
- The input parameters of the ISF process control the quality of the product and cycle time.
- ISF parameters are classified into workpiece parameters, machine parameters and process parameters.
- The right choice of tool-path largely improves the formability, surface finish and dimensional accuracy of the components formed using the ISF process. Furthermore, the processing time of components can be greatly reduced by choosing an appropriate tool-path.
- The tool-path is classified into profile and helical. The profile tool-path is classified into several sub-categories based on the tool movement.
- The helical tool-path provides uniform forming-forces and produces a better surface quality of the components.
- The responses of the ISF process such as forming-force, formability, dimensional accuracy, thinning limit and surface roughness, etc., determine the criteria of setting up the process parameters for various applications.
- A process engineer can alter the ISF process parameters to control the various responses of the process to make it suitable for forming the desired component.
- The force components $Fx$, $Fy$ and $Fz$ along the $x$, $y$ and $z$ directions, respectively, can be measured with the help of a force dynamometer or load cell.

- The tool-path is an important parameter of the ISF process that affects the forming-force.
- The angular step-down tool-path is the worst type of tool-path because it requires a higher amount of forces to form the components as compared to vertical step down and helical tool-path.
- Forming-force increases with the increase in tensile strength of the workpiece material.
- The greater strain-hardening exponent of the material increases the resistance of the material to plastic deformation. Such materials require greater forming-forces to deform the sheet.
- The gradient of the force curve after peak values can be considered and utilized as a hidden variable for modifying process parameters by continuous comparison of the critical stress value and instant stress value for safe execution of the forming operation.
- The formability of the ISF process is much higher than that of conventional forming processes due to the nature of the mechanism of deformation in this die-less process.
- The formability of the ISF process can be explained by the maximum forming depth, maximum wall angle and FLDs.
- The forming limits of the ISF process can be termed as FFL. The FFL line is a straight line with a negative slope of unity in the positive major and minor strain region in FLD.
- The actual shape of the formed component after unloading from fixture deviates from the desired CAD model due to elastic strains known as the springback effect. This deviation of the final component from the desired one or the geometrical inaccuracy can be in the form of sheet bending near the clamping area of the workpiece, wall deviation or pillow effects.
- The use of a backing plate prevents sheet-bending near the clamping area.
- The prediction of springback in formed parts could become the key to accuracy of the ISF process.
- Process accuracy can be enhanced and deviations from target shape can be minimized by reducing springback effects using the following strategies: Applying HA-ISF, online control of process parameters, optimizing tool-paths and input parameters.
- Thickness reduction, thinning limit and thickness distribution can be considered as important criteria to predict the feasibility of the ISF process.
- Sheet thinning can be measured by different methods, viz., micrometer, CMM and 3D scanners.
- Strain and thinning are higher for steeper wall angles and lead to the fracture of the component if the thinning limit is reached.
- Forming time can be greatly reduced by optimizing tool-paths and applying high permissible feed rates.
- The breakeven point of the ISF process can be a few parts as it does not require the manufacturing of dedicated dies.
- Different biocompatible materials like titanium grade 1 and grade 2, polycaprolactone and nano-polymer composites can be formed by using ISF to

produce various medical implants like ankle support, cranial plate, facial implants, knee prosthesis and so on.

- ISF has capabilities to form large-size double curved surfaces, and this is in addition to the traditional architectural lining methods like rubber forming.
- Some applications of ISF in the automotive and aerospace fields are engine heat shields, car fenders, airfoils, taillight brackets of cars, covers of motorcycle exhaust pipes and so on.
- Some challenging bottlenecks to implementing the ISF process on an industrial scale are the following: High forming time, lower accuracy, difficulty in formation of vertical walls, poor surface finish, excessive material thinning and the lack of mathematical models and analytical methods to predict the parameters of the process like forming-force, formability, surface roughness, strains, thinning limits and so on.
- ISF is a green choice in the field of rapid prototyping and for small batch size production. It saves as much as 10% material compared to conventional sheet forming processes and also saves cost of tooling (die and punch), as it directly avoids the use of forming dies and punches.

There have been remarkable developments and research in ISF in the last two decades, but a lot more needs to be done to make ISF processes commercially viable for industry. There is significant scope for using new advancements like industry 4.0 technologies, and ISF remains to be perfected. Intensive modeling and investigations are required to be carried out for ISF processes before they will be accepted commercially by industry.

## 8.3 SCOPE OF FUTURE WORK

A large number of application areas can be benefited by the flexible forming capabilities of ISF. Nevertheless, the process is waiting for full implementation on an industrial scale. Manufacturing of complex and robust shapes with relative ease can make the ISF process a success in the future. The future possibilities of this die-less forming technology are extremely promising. Systematic process planning, tool-path optimization, compensating unwanted deformations, advanced feature recognition techniques, forming reliable and robust communication interfaces, flexible hardware, adaptive control methods, the ability to form newer materials and a wider range of materials, complex shapes with ease and accuracy, removing barriers on sizes of components and so on are the dominant key issues to be focused on in future. The optimization of process variants and process factors can significantly enhance the performance of the process. Improving the geometrical accuracy and formability of the process is another key research area. These evolutions would directly lead to many technical and novel developments in the field of ISF technologies. To fabricate sandwich panels by ISF process is another emerging area for future work.

To conclude, we can pronounce that ISF techniques are novel techniques that are based on forming the material layer by layer locally, to generate the desired shape. It has a long way to go for future research, development and commercialization. The authors' attempt is modest and unpretentious with an aim to manifest a few pearls of knowledge and consciousness for the technical fraternity.

# Appendix A: List of US, Japanese (JP), European (EP) and International (WO) Patents Related to ISF

| S. No. | Country-Patent Number | Patent File Date<br>Patent Issue Date | Issued to/Filed by | Title of Patent |
|---|---|---|---|---|
| 1 | US 3342051 | August 10, 1964<br>September 19, 1967 | Edward Leszak | Apparatus and process for incremental dieless forming |
| 2 | US 3316745 | February 8, 1965<br>May 2, 1967 | Walter G. Berghahn, Lee, and George F. Murray, Jr. (General Electric Company) | Method of dieless forming surfaces of revolution |
| 3 | JP 10-180365, P3445988 | December 12, 1996<br>July 7, 1998 | Hideo Iseki, Takurou Uede, Naoki Simode | Incremental bulging machine and the forming tool unit |
| 4 | JP 07-132329 | November 11, 1993<br>May 23, 1995 | Shigeo Matsubara | Method and device for forming metal plate |
| 5 | JP 09-010855 | June 30, 1995<br>January 14, 1997 | Takahashi Yoshida, Shigeo Matsubara (Matsushita Electric Ltd. | Method and device for forming sheet |
| 6 | JP 09-085355 | July 18, 1996<br>March 31, 1997 | Akio Shima, Takenao Yoshikawa, Keiichi Nakamura, Yoichi Suda, Shoji Suzuki (Hitachi Ltd.) | Formation of successively expanding metallic plate and apparatus therefor |
| 7 | JP 10-137858 | November 01, 1996<br>May 26, 1998 | Katsutoshi Ueno, Fuminori Matsuda, Tatsuya Nagata, Kazuhiko Yamamoto, Atsunobu Murata, Fumio Nonioyama (Toyota Motor Corp., Taiho Seiki Co. Ltd., Toyota CRDL Inc.) | Incremental forming method |
| 8 | JP 10-296345 | April 25, 1997<br>November 10, 1998 | Kenichi Waratani, Takaneo Yoshikawa, Keiichi Nakamura, Akio Shima, Hidenori Endo, Yasuhiko Tajima, Shoji Suzuki, Takashi Misdeki (Hitachi Ltd.) | Method for continuously and successively bulging metal sheet and device therefor |

*(Continued)*

| S. No. | Country-Patent Number | Patent File Date<br>Patent Issue Date | Issued to/Filed by | Title of Patent |
|---|---|---|---|---|
| 9 | JP 10-314855 | May 22, 1997<br>December 02, 1998 | Katsutoshi Ueno, Fuminori Matsuda, Tatsuya Nagata, Kazuhiko Yamamoto, Atsunobu Murata, Fumio Nonioyama (Toyota Motor Corp., Toyota CRDLInc, Taiho Seiki Co. Ltd.) | Male die forming method of metal sheet |
| 10 | JP 11-285741 | April 03, 1998<br>October 19, 1999 | Izumi Ochiai, Kazuo Momoiyama, Takeshi Onawa, Shoji Suzuki, Takashi Miseki (Hitachi Ltd., Hitachi Tochigi Material:KK) | Formation of metallic plate and forming tool |
| 11 | JP 2000-153313 | November 16, 1998<br>June 06, 2000 | Atsunobu Murata, Fuminori Matsuda (Toyota CRDL Inc., Toyota Motor Corp.) | Incremental pressing and forming device |
| 12 | JP 2002-102944 | September 25, 2000<br>April 09, 2002 | Ichiro Kono, Kenichi Hayashi, Koji Fujita, Akira Oba, Hideo Meguri (Honda Motor Co. Ltd.) | Incremental stretch forming method |
| 13 | EP 0970764 A1 | January 29, 1999<br>January 12, 2000 | Shigeo Matsubara, Hiroyuki Amino, Susumu Aoyama, Yan Lu (Amino Corporation) | Apparatus for dieless forming plate materials |
| 14 | US 6216508 B1 | January 29, 1999<br>April 17, 2001 | Shigeo Matsubara, Hiroyuli Amino, Susumu Aoyama, Yan Lu (Amino Corporation) | Apparatus for dieless forming plate materials |
| 15 | US 6532786 B1 | April 19, 2001<br>March 18, 2003 | Clint Allen Luttgeharm (D-J Engineering, Inc.) | Numerically controlled forming method |
| 16 | US 6748780 B1 | March 13, 2003<br>June 15, 2004 | Clint Allen Luttgeharm (D-J Engineering, Inc.) | Numerically controlled forming method |
| 17 | EP 1147832 A2 | April 17, 2001<br>October 24, 2001 | Norihisa Okada, Wataru Kawamura (Hitachi Ltd.) | Incremental forming method and apparatus for the same |

*(Continued)*

| S. No. | Country-Patent Number | Patent File Date<br>Patent Issue Date | Issued to/Filed by | Title of Patent |
|---|---|---|---|---|
| 18 | US 6561002 B2 | April 11, 2001<br>May 13, 2003 | Norihisa Okada, Wataru Kawamura (Hitachi Ltd.) | Incremental forming method and apparatus for the same |
| 19 | EP 1462189 A1 | August 21, 2003<br>September 29, 2004 | Norihisa Okada, Gen Ro, Yohji Suzuki (Hitachi Ltd., Amino Corporation) | Method and apparatus for incremental forming |
| 20 | US 6971256 B2 | August 19, 2003<br>December 6, 2005 | Norishia Okada, Gen Ro, Yohji Suzuki (Hitachi Ltd. and Amino Corporation) | Method and apparatus for incremental forming |
| 21 | EP 1899089 A2 | April 21, 2006<br>October 26, 2006 | Bart Callebaut, Joost Duflou, Johan Verbert (K.U. Leuven Research and Development) | Asymmetric Incremental Sheet Forming system |
| 22 | US 0158805 A1 | April 21, 2006<br>June 25, 2009 | Bart Callebaut, Joost Duflou, Johan Verbert (K.U. Leuven Research and Development) | Asymmetric Incremental Sheet Forming system |
| 23 | EP 1462190 A1 | August 28, 2003<br>September 29, 2004 | Norihisa Okada, Gen Ro, Yohji Suzuki, Masatoshi Nara (Hitachi Ltd., Amino Corporation) | Method and apparatus for incremental forming |
| 24 | US 0148997 A1 | January 30, 2003<br>August 5, 2004 | Hiroyuki Amino, Shigeo Matsubara, Yan Lu | Shaping method and apparatus of thin metal sheet |
| 25 | EP 1731238 A1 | June 06, 2006<br>December 13, 2006 | Hiroyuki Amino, Yan Lu, Shigeo Matsubara (Amino Corporation) | Method and apparatus for numerically controlled, in at least two steps die-less sheet metal forming |
| 26 | US 0272378A1 | June 5, 2006<br>December 7, 2006 | Hiroyuki Amino, Shigeo Matsubara, Yan Lu | Method and apparatus for forming sheet metal |

*(Continued)*

| S. No. | Country-Patent Number | Patent File Date / Patent Issue Date | Issued to/Filed by | Title of Patent |
|---|---|---|---|---|
| 27 | EP 1560668 A1 | October 03, 2003<br>April 15, 2004 | TimoTuominen (TwincamOy) | Method and apparatus for forming three-dimensional shapes in a sheet metal |
| 28 | EP 1977842 A1 | January 19, 2007<br>October 08, 2008 | Pedro Pablo Rodriguez Gutierrez, Maria Asuncion Rivero Rastero, Elixabete Maidagan Onandia (Fundacion Fatronik) | Machine for shaping sheet metal and shaping method |
| 29 | US 8021501 B2 | August 19, 2008<br>September 20, 2011 | John Roth (Penn State Research Foundation, US) | Single point incremental forming of metallic materials using applied direct current |
| 30 | US 0226272 A1 | March 6, 2009<br>September 10, 2009 | Kevin Scott Smith, Bethany A. Woody, John C. Ziegert, Jian Cao (University of North Carolina at Charlotte, Clemson University, and Northwestern University) | Deformation machining systems and methods |
| 31 | US 2010/0199742 A1 | February 11, 2009<br>August 12, 2010 | Carl Frederick Johnson, Vijitha Senaka Kiridena, Feng Ren, Zhiyong Cedric Xia (Ford Global Technologies, LLC, US) | System and method for incrementally forming a workpiece |
| 32 | US 8302442 B2 | July 29, 2010<br>November 06, 2012 | Global Technologies, LLC, US) | Method of incrementally forming a workpiece |
| 33 | US 2012/0024034 A1 | July 27, 2010<br>February 02, 2012 | Feng Ren, Zhiyong Cedric Xia (Ford Global Technologies, LLC, US) | Method to improve geometrical accuracy of an incrementally formed workpiece |
| 34 | US 2012/0011915 A1 | July 15, 2010<br>January 19, 2012 | Vijitha Senaka Kiridena, Feng Ren, Zhiyong Cedric Xia (Ford Global Technologies, LLC, US) | Method of incremental forming with successive wrap surfaces |

*(Continued)*

| S. No. | Country-Patent Number | Patent File Date<br>Patent Issue Date | Issued to/Filed by | Title of Patent |
|---|---|---|---|---|
| 35 | WO 2012/030835 | August 30, 2011<br>March 08, 2012 | John Roth (The Penn State Research Foundation, US and Northwestern University, US) | Electrical-assisted double side incremental forming and processes thereof |
| 36 | US 2013/0103177 A1 | October 17, 2012<br>April 25, 2013 | Jian Cao, Rajiv Malhotra (Northwestern University, US) | System and method for accumulative double sided incremental forming |
| 37 | US 10010920 B2 | June 10, 2014<br>July 03, 2018 | Feng Ren, Zhiyong Cedric Xia (Ford Global Technologies, LLC, US) | Method to improve geometrical accuracy of an incrementally formed workpiece |
| 38 | US0257909 A1 | April 08. 2009<br>October 14, 2010 | C.S. Huskamp, J.B. Castle, and K.G. Waymack (The Boeing Company, US) | Method and apparatus for reducing force needed to form a shape from a sheet metal, |
| 39 | US 9682418 B1 | June 18, 2009<br>June 20, 2017 | Keith A. Young, Richard J. Lederich, James B. Castle, and Christopher S. Huskamp | Method and apparatus for incremental sheet forming |
| 40 | US 0042683 A1 | August 10, 2012<br>February 13, 2014 | Vijitha Senaka Kiridena, Zhiyong Cedric Xia (Ford Global Technologies, LLC, US) | Fixture assembly for forming prototype parts on an incremental forming machine |
| 41 | US 9522420 B2 | April 02, 2013<br>December 20, 2016 | Kiyoshi Nonomura, Koji Kurozumi, Koji Inoue (Toyota Jidosha Kabushiki Kaisha, Japan) | Incremental forming method |
| 42 | US 9676019 B2 | June 13, 2013<br>June 13, 2017 | The Boeing Company, US | Determining part orientation for incremental sheet forming |
| 43 | US 10144048 B2 | November 19, 2014<br>December 04, 2018 | Vijitha Senaka Kiridena, Zhiyong Cedric Xia, Feng Ren (Ford Global Technologies, LLC, US) | High stiffness and high access forming tool for incremental sheet forming |

*(Continued)*

| S. No. | Country-Patent Number | Patent File Date<br>Patent Issue Date | Issued to/Filed by | Title of Patent |
|---|---|---|---|---|
| 44 | US 0082498A1 | November 30, 2015<br>March 24, 2016 | Michael Beltran, Jian Cao, John T. Roth (The Penn State Research Foundation, US and Northwestern University, US) | System and method for incremental forming |
| 45 | EP 3078433 A1 | March 30, 2016<br>October 12, 2016 | John Huntley Belk, Om Prakash (The Boeing Company, US) | Method and system for incremental sheet forming of tailored blanks |
| 46 | US 0113432 A1 | October 25, 2016<br>April 26, 2018 | Justin L. Peters (The Boeing Company, US) | Methods and systems for generating a continuous helical tool path for parts having multiple central axes |
| 47 | US 10162329 B2 | February 03, 2017<br>December 25, 2018 | Ebot E. Ndip-Agbor, Jian Cao (Northwestern University, US) | Automated toolpath generation method for double sided incremental forming |
| 48 | US 0030585 A1 | July 26, 2017<br>January 31, 2019 | Andrey Ilinich, S. George Luckey, Alan John Gillard, Vijitha Kiridena (Ford Global Technologies, US) | Method to reduce tool marks in incremental forming |

# Index